Fundamental Processes in Ecology

An Earth Systems Approach

David M. Wilkinson
Liverpool John Moores University, UK

OXFORD
UNIVERSITY PRESS

OXFORD

UNIVERSITY PRESS

Great Clarendon Street, Oxford OX2 6DP

Oxford University Press is a department of the University of Oxford.
It furthers the University's objective of excellence in research, scholarship,
and education by publishing worldwide in

Oxford New York

Auckland Cape Town Dar es Salaam Hong Kong Karachi
Kuala Lumpur Madrid Melbourne Mexico City Nairobi
New Delhi Shanghai Taipei Toronto

With offices in

Argentina Austria Brazil Chile Czech Republic France Greece
Guatemala Hungary Italy Japan Poland Portugal Singapore
South Korea Switzerland Thailand Turkey Ukraine Vietnam

Oxford is a registered trade mark of Oxford University Press
in the UK and in certain other countries

Published in the United States
by Oxford University Press Inc., New York

British Library Cataloguing in Publication Data

Data available

Library of Congress Cataloging in Publication Data

Wilkinson, David M., 1963–
    Fundamental processes in ecology: an earth systems approach / David
    M. Wilkinson.
        p. cm.
    ISBN-13: 978–0–19–856846–9 (alk. paper)
    1. Ecology.   2. Earth sciences.   I. Title.
    QH541.W455 2006
    577—dc22   2006012753

Typeset by Newgen Imaging Systems (P) Ltd., Chennai, India
Printed in Great Britain
on acid-free paper by
Biddles Ltd, King's Lynn, Norfolk

ISBN 978–0–19–922906–2 (Pbk.)   978–0–19–856846–9 (Hbk.)

10 9 8 7 6 5 4 3 2 1

The answers to nearly all the major philosophical questions are either found in or illuminated by the science of life, especially ecology, whose stated goal is the elucidation of the relationship of organisms to environment.

Lynn Margulis (in Margulis and Sagan, 1997, p. 311).

Preface

A work of synthesis is not only difficult, it is also not part of the pattern of a career in modern science.

Nisbet (1991, p. xvi)

This volume is a book-length expansion of a paper originally published in *Biological Reviews* during 2003. That paper summarized an approach to thinking about academic ecology at the scale of the whole Earth based on the following astrobiological thought experiment; 'for any planet with carbon-based life, which persists over geological time scales, what are the minimum set of ecological processes that must be present?' In the context of all the changes that we are making to our planet, such an Earth Systems approach to ecology appears timely—if only to highlight what we don't currently understand. In addition, I believe that there are great benefits in rethinking very fundamental ideas from a novel perspective and asking questions about some of the traditional emphases of ecology: such as why are so many of our journals filled with huge flocks of birds when microorganisms are so much more central to the things we need to understand?

Attempting to put ecology into this much wider context requires the synthesis of information from biology, chemistry, physics, geology, and astronomy; few readers can be expected to be fluent in the language of all these parts of science so I have included an extensive glossary (which also summarizes the geological time scale). As the epigraph to this preface suggests, such a work of synthesis was considered both difficult and unfashionable by Euan Nisbet at the start of the 1990s when he wrote his own, more applied, book on the Earth—today it is no less difficult and probably even more unfashionable, at least within British science. The problem is that such synthetic work requires no expensive laboratory or groups of research students or post Docs and as such it does not generate the large research grants, rich in overheads, for the scientist's University or Institute. The influence of such work is also very difficult to quantify, at least on a short time scale, this makes it almost useless to the accountants of research 'quality'. As such, many academic administrators consider this kind of work completely pointless—I am lucky that at least some of my senior colleagues at Liverpool John Moores University have been tolerant of my overhead-free theoretical work. I would probably have been in danger of losing my job at many British Universities for following this approach to research.

I don't expect anyone to agree with all the ideas and suggestions in this book—that includes myself as I will have no doubt changed my views on at least a few points of detail before publication! I will consider the book successful if it provokes people into reconsidering the core ideas of ecology and hopefully in so doing causing some of them to generate new and interesting ideas themselves. At the very least the reference list should be a useful way into a diverse literature, both for ecologists interested in the wider context of their subject and for scientists from other disciplines interested in what ecology can offer their subject.

Many people have contributed towards the writing of this book. First my science teachers must take some responsibility for my approach (hopefully this will not dismay them!): I would particular like to highlight the role of my father, Lionel Wilkinson in my science education. In addition, both Humphrey Smith and David Keen were influential at Coventry Polytechnic while Brian Huntley and John Coulson were especially important while I was at the University of Durham. Fred Slater (curator of Cardiff University's biological field centre) was influential in giving me the opportunity to 'play' with a wide range of taxonomic groups in the late 1980s. It gives me great pleasure that the science discussions with my father are still ongoing and that in recent years I have returned to collaborative work with Humphrey Smith in his 'retirement'. Yrjo Haila, David Schwartzman, and Tyler Volk provided helpful comments on the paper that was the forerunner to this book (Wilkinson, 2003). Jim Lovelock, Hannah O'Regan (my wife), and Tim Lenton provided comments on early drafts of the majority of chapters; Tim in particular provided many detailed suggestions in spite of his own heavy work load. Tom Sherratt, Andy Young, and Ian Sherman also provided invaluable comments on some of the chapters. Tom has been an especially important collaborator in theoretical work since the end of the 1990s, some of our joint work being facilitated by financial support from The Royal Society of Britain. Three of the graphs used in this book have been redrawn from other authors' work; I thank Jim Lovelock, Graeme Ruxton, Tom Sherratt, Mike Speed, Bland Finlay, and Tom Fenchel for permission to use their work in this way. My former research student Steve Davis helped with the production of Figs 4.1 and 4.5; the photographs are my own.

David M. Wilkinson

Liverpool
January 2006

Contents

Part I *Introduction*

In fact, the conditions necessary for life on the earth have not been 'naturally' there but have been shaped by life itself.

Haila (1999a, p. 338)

1. *Introducing the thought experiment*

1.1. **The entangled bank**

The subject matter of ecology appears confusingly complex, one of the most famous images of this complexity in the scientific literature being Charles Darwin's description of an entangled bank (Fig. 1.1). In the closing pages of *The Origin of Species* he wrote, 'It is interesting to contemplate an entangled bank, clothed with many plants of many kinds, with birds singing on the bushes, with various insects flitting about and worms crawling through the damp earth.' (Darwin, 1859, p. 489). Some years ago in an essay in the journal *Oikos* I argued that Darwin's image ignored many of the most important parts of the system, such as the microorganisms in the soil and the mycorrhizal fungi in the plant roots (Wilkinson, 1998). Indeed, while Darwin stressed the animals (birds, insects, and worms), it is the plants, fungi, and, most importantly, the microbes which are involved in the majority of ecosystem services (Fig. 1.2). Indeed, for most of the history of life on Earth ecology was entirely microbial, although today microbes are surprisingly rare in our general ecology textbooks. In the 1998 essay I wrote, 'To a first approximation the animals are of no importance to the functioning of the system.' Possibly I was slightly too hard on the animals and their lack of functional importance. As Joel Cohen (pers. comm.) has pointed out to me, although animals may not be required for most of the important ecosystem processes, they should be regarded as potentially important catalysts, as their presence may greatly speed up the *rate* of these processes. For example, Darwin's worms are probably catalysing the microbial breakdown of leaf litter.

Ecological systems such as Darwin's entangled bank (composed of myriad species of microbes, fungi, plants, and animals; all interacting with each other and their abiotic environment) are clearly difficult to understand and challenging to explain to students or the wider public. It is instructive to ask, *how has academic ecology attempted to order this complexity?* I believe the answer to this question can be most easily seen by looking at university-level general ecology

Fig. 1.1: Archaeological evidence suggests that hedgerows have been a part of the British landscape since at least late prehistoric times—here a hedge marks out one side of a 'strip field' of probable medieval date behind the village of Great Asby in Cumbria, northwest England (see Clare, 1996). In some parts of Britain they are traditionally established on hedgebanks, which are constructed of earth or stones, these are common in southwest England but also found much more widely across the country. Before metalled road surfaces many British roads formed eroded troughs often edged with hedges on the tops of their banks, so-called holloways (Rackham, 1986; Muir, 2004). In summer these banks can be alive with life, Darwin probably had similar hedges in mind when he constructed his famous metaphor of an 'entangled bank'—which he shortened to a 'tangled bank' in later editions of *The Origin of Species*.

texts. As Stephen J. Gould (2002, p. 576) wrote in his *magnum opus* 'yes, textbooks truly oversimplify their subjects, but textbooks also present the central tenets of a field without subtlety or apology—and we can grasp thereby what each generation of neophytes first imbibes as the essence of the field . . . I have long felt that surveys of textbooks offer our best guide to the central convictions of an era'.

1.2. **The entity approach**

How do textbooks organize ecology? In most cases as a hierarchy of entities (I am indebted to Haila (1999b) for the term 'entities' in this context). There appears to be a reasonable consensus about how to classify these ecological entities in a hierarchical manner: going from genes through individuals, populations,

Fig. 1.2: A biological soil crust (cryptogamic soil) in the Utah Desert, USA. Such crusts often have filamentous cyanobacteria as an important part of their structure, along with mosses and lichens. These crusts are very important in many arid and semi-arid systems around the world; especially in controlling hydrological aspects of the soil and preventing soil erosion (Lange *et al.*, 1992). They provide a rare example of a system where the ecological importance of microbes is clearly visible to the unaided human eye. Mainstream ecology has tended to concentrate on macroscopic organisms (Fenchel, 1992; Wilkinson, 1998), however, there is now some sign of a slight increase in microbial papers in several ecological journals, although this work has yet to feature in many ecology textbooks. Microbial studies are crucial for the development of an Earth Systems ecology, as such they feature prominently in this book.

species, communities, ecosystems to the biosphere (e.g. Colinvaux, 1993; Smith and Smith, 1998; Krebs, 2001; Stiling, 2002; Begon *et al.*, 2006). A hierarchical approach, from population to community ecology, was also used to structure the *Principles of Animal Ecology* (Allee *et al.*, 1949), one of the key textbooks in the mid twentieth century. This approach has an even longer pedigree in plant ecology with an important early text (Warming, 1909) using a different entity-based strategy, being arranged around various plant communities. Indeed an entity approach is an obvious way of organizing natural history into species, habitats, biomes, and so forth. An early exception to these approaches was Charles Elton's classic text *Animal Ecology* from the 1920s (Elton, 1927), which he organized, at least in part, around concepts; such as 'succession', 'parasites', and 'dispersal'.

An entity-based approach has great strengths in describing systems. It is probably also inspired by a strongly reductionist tradition which believes that the lower levels in a hierarchy contain all the information needed to understand the higher levels. Reduction has been so successful in tackling many problems in the past that the philosopher Mary Midgley (2001) has argued that there has been an unfortunate tendency for some scientists to think that it is not only necessary but also sufficient to explain any scientific problem. In this view community ecology is just population ecology writ large, so that an understanding of communities will provide everything required to understand the biosphere. An interesting recent example of this approach is by Allen *et al.* (2005) who describe a 'bottom-up' model which attempts to make predictions about the global carbon cycle based on the effects of body size and temperature on individual organisms. However such bottom-up, reductionist, approaches may have limits when faced with all the complexities of the entangled bank. In studying a complex system it is often its organization that is most important. Ernst Mayr, in his last book, illustrated this point with the following physiological example: 'No one would be able to infer the structure and function of a kidney even if given a complete catalog of all the molecules of which it is composed' (Mayr, 2004, p. 72). In an ecological context, Lawton (1999) has argued that the many complex contingencies in ecological systems may limit such an approach, forcing us to rely less on reduction and experimental manipulations, especially at the level of community ecology.

1.3. **A process-based approach**

An obvious alternative to the hierarchical entity approach would be to emphasize processes, especially if the goal is conceptual understanding rather than a narrative description of the natural world. Such an approach immediately raises a crucial question: what are the fundamental processes in ecology? While many authors appear to agree on the broad outline of an entity approach (genes to biosphere), no such consensus is available for ecological processes. This book is a provisional attempt to address this difficult question.

The type of approach used is important as it can govern the kinds of ecological questions a researcher asks. Consider peatland systems, the entity approach suggests questions about the number of different peatland types. Such questions date back to Linnaeus in the eighteenth century, who appears to have been the first person to publish lists of plant species from different types of bogs (Du Rietz, 1957). Much more recently the British National Vegetation Classification has defined 38 peatland plant communities to be found in Great Britain (Rodwell, 1991). A process-based approach to peatlands would more naturally lead to very different questions, for example about their role in carbon sequestration and its climatic implications (e.g. Klinger *et al.*, 1996; Clymo *et al.*, 1998). A recent analysis of ecological research papers published over the last 25 years suggests that there is a growing increase in studies of processes (Nobis and Wohlgemuth, 2004). If the approach taken can affect the question asked then it can clearly affect our understanding of the Earth. If a more processed-based approach is to be considered then it becomes important to develop a reasonably rigorous way of defining key, or fundamental, ecological processes.

One possibility would be to ask ecologists what processes they consider important. In the run up to the 75th anniversary of the British Ecological Society a survey of its membership asked them to rank a list of ecological concepts in order of their importance (Cherrett, 1989). The resulting top five concepts were; 'the ecosystem', 'succession', 'energy flow', 'conservation of resources', and 'competition'. It is interesting that some of these overlap with those used by Elton (1927) in his early attempt at a concept-based ecology text. However, Cherrett's concepts differ from what I mean in this book by 'fundamental processes'; for example, 'conservation of resources' only applies to a planet populated with intelligent organisms that can plan ahead. Other concepts discussed by Cherrett, such as the idea of nature reserve management or maximum stainable yields (by harvesting humans) are also clearly not fundamental ecological processes but important applied concepts for a planet with intelligent life.

An alternative strategy would be to attempt to approach fundamental ecological processes experimentally using mesocosms. However, it is unclear if such systems are large enough (and they clearly operate over an unrealistically short time scale) to answer such big conceptual questions. The largest of these experimental systems has been the 1.3 ha Biosphere 2 closed-environment facility in Arizona, USA. However, so far, the main theoretical contribution of this mesocosm has been to illustrate the difficulties in maintaining stable ecological systems, even on a year-to-year basis (Cohen and Tilman, 1996). Therefore, even the largest mesocosm apparently does not provide a realistic system for experimental study of many major ecological processes.

An alternative approach is to use the idea of thought experiments. Although these are commonly used in philosophy and theoretical physics they are less common in ecology and the environmental sciences. However, thought experiments have a long history of use in addressing problems where direct experiment is difficult or impossible (see Sorensen (1991), for a useful short introduction). Whereas a conventional experiment usually adds new data from the 'real world', a thought experiment sets up imaginary scenarios with the intention of adding to understanding by investigating the experiment's internal consistency or compatibility with what is already known.

Although commoner in other subject areas, thought experiments are not unknown in ecology. A well-known example is W.D. Hamilton's (1971) paper *'Geometry for the selfish herd'*. This paper started thus: 'Imagine a circular lily pond. Imagine that the pond shelters a colony of frogs and a water-snake'. Hamilton famously went on to use this simplified imaginary scenario to test the logic of ideas about the anti-predator advantages of living in groups. This paper is now considered a classic of evolutionary ecology, showing that thought experiments can sometimes be powerful tools for thinking about ecological ideas.

This book is structured around the following thought experiment. 'For any planet with carbon-based life, which persists over geological time-scales, what is the minimum set of ecological processes that must be present?' By limiting myself to considering carbon-based life located on a planet many possible life forms are excluded. For example, the astronomer Fred Hoyle invented extraordinary intelligent interstellar clouds in his novel *The Black Cloud* (Hoyle, 1957). This is one of the most interesting alien life forms in science fiction as it relies on neither Earth-type biology or a planetary habitat. Writing as a scientist he also argued for carbon-based life living in comets, although many have viewed this as another aspect of Hoyle's science fiction writing (Hoyle and Wickramasinghe, (1999a); Wickramasinghe (2005) for a more accessible autobiographical account of this work). It is possible that Hoyle himself may have become less convinced by some of the more extreme versions of these ideas towards the end of his life (Gregory, 2005). Clearly it is quite possible to think of potential biological systems and ecologies which fall outside the carbon-based, planet-based, boundaries of the discussion in this book. However to be useful a thought experiment needs constraints to limit the scope of the speculation. The constraints in my thought experiment seem reasonable as Pace (2001) has outlined arguments which suggest that any life in the universe is likely to be based on organic chemistry (Box 1.1). In addition, we have studied one example (the Earth), and similar conditions may exist around many main sequence stars (Kasting *et al.*, 1993; Franck *et al.*, 2004). Indeed very un-Earth like places may allow life to survive if a heat source allows life access to water in a form which allows solute diffusion (Pace, 2001). Possible examples include heat from internal planetary sources or 'impact

---

Box 1.1: **The universal nature of biochemistry**

Norman Pace (2001) has outlined a number of reasons for thinking that any life in the universe will be based on organic chemistry. Here I briefly summarise his main arguments; see his original essay for more details.

Only two atoms are known to serve as the backbone of molecules large enough to carry biological information; carbon and silicon. Carbon is much more likely because

1. Carbon readily forms chemical bonds with a wider range of other atoms than silicon.

2. The electronic properties of carbon allow the formation of double and triple bonds with other atoms, while silicon does not readily form such bonds. These bonds allow for the capture and storage of energy from the environment, a crucial ability for any form of life.

Which organic chemicals?

Within the 'chemical space' of possible organic chemicals, life on Earth appears to use a rather limited selection. This may be an example of historical accident (in which case different organic chemicals may be used by any life elsewhere in the universe) or may be due to chemical constraints which could apply to all life forms. For example the requirement of solubility in water may be a reason why many of the smaller organic molecules used by life on Earth are derivatives of simple carboxylic acids and organic amines (Dobson, 2004); however water is not the only solvent for organic chemistry so there is at least an outside chance that is not necessary for carbon based life (Ball, 2005a).

---

oases' producing liquid water on an otherwise frozen astronomical body, due to an asteroid or a comet impact (O'Brien *et al.*, 2005). Although questions about the possibility of life on planets other than the Earth are fascinating, my main intention with this book is to use an astrobiological perspective as a means of thinking about the fundamental processes of ecology on Earth. In the context of all the changes we are making to our planet, an attempt to recast academic ecology from such an Earth Systems' perspective appears worthwhile.

1.4. **Gaian effect**

In the case of the Earth, life is not a new phenomenon but has existed for much of its history. Our planet is around 4,600 million years old and many authorities consider it likely that life was present by at least 3,500 million years ago, indeed this figure has been cited in many review articles and textbooks in recent decades (e.g. Brasier, 1979; Raven and Johnson, 1999). The very earliest dates don't rely on conventional fossils but on the interpretation of carbon isotopes recovered from rocks around 3.8 billion years old in Greenland (Rosing, 1999; Nisbet and Sleep, 2001); recently these results have become very controversial, with doubts being raised both over the evidence for life and the dating of the rocks (Lepland *et al.*, 2005; Moorbath, 2005). The oldest fossil microorganisms are considered by many to come from the Apex Chert in Australia and are in excess of 3,465 million years old (Schopf, 1999), and are responsible for the textbook dates of 3.5 billion years for the first life. However, these too now appear less certain, having been controversially reinterpreted as secondary artefacts of amorphous graphite formed around hydrothermal vents (Brasier *et al.*, 2002). Whatever the nature of these 'fossils', Brasier *et al.* (2002) conceded that 'carbon isotopic values from the graphite cherts imply a significant biological contribution to the carbon cycle' at the time, so even if Schopf's microfossils turn out to be artefacts there is still good circumstantial evidence for life at this date. Five hundred million years later, by around 3 billion years ago, there is abundant evidence for life, with relatively uncontroversial fossil evidence for microbial life on Earth such as stromatolites described from several different locations (Schopf, 1999; Nisbet, 2001; Nisbet and Sleep, 2001). Thus the age of the first evidence for life is currently in flux and as such this paragraph is likely to be one of the first sections of this book to become seriously out of date! Currently, it seems safe to claim evidence for life at 3 billion years, good circumstantial evidence at 3.5 billion and a real possibility that life on Earth is older, although currently we are lacking good evidence for this.

Prior to 3.8 billion years ago, conditions for life on Earth would have been very challenging. Evidence from our Moon (where the relevant impact craters are preserved unlike the situation on Earth) suggests that the early history of the Earth would have been marked by large asteroid impacts, which could have made any surface life impossible, this would rule out photosynthetic organisms which by definition must live on the surface of a planet to gain access to life (Sleep *et al.*, 1989). If there was life present during this period, there appears to be two main ways in which it could have survived. The first possibility would be by living deep within the Earth's rocks, the 'deep hot biosphere' popularized by Thomas Gold (1999). The second possibility is life surviving off our planet. The most widely discussed possibility being the colonization, or recolonization, of the Earth from Mars via microbes transported in meteorites originating from planetary surfaces (Davies, 1998). An interesting variation on this second idea is that during a large asteroid impact on Earth microorganisms could be ejected into space inside rocks, which later fall back onto the now sterile Earth so reseeding our planet (Wells *et al.*, 2003). Clearly these various possibilities are not mutually exclusive.

Whatever the date of the first life on Earth it is clear that it has existed for at least 3 billion years and this raises an interesting problem, given all the things that could have happened to cause its extinction, why has it survived so long? The range of potential disasters is wide, including the impact of a large meteorite (objects in excess of 500 km in diameter could potentially vaporize the Earth's oceans (Nisbet and Sleep, 2001)), or global ice ages, extreme versions leading to 'Snowball Earth'. More complex life of the kind that most ecology textbooks concentrate on tends to be less tolerant than prokaryotes (Table 1.1 illustrates this for the upper-temperature limit for growth) and as such it could be exterminated by less extreme events than those described above.

Although ecology has traditionally considered the abiotic environment as an unchanging stage on which organisms act out their ecological relationships, there is another fascinating possibility: that feedbacks between the organisms and their environment have helped to maintain habitable conditions on Earth. This is the Gaia hypothesis, developed by James Lovelock and Lynn Margulis in the late 1960s and 1970s. (For more recent descriptions of this idea

Table 1.1: Approximate upper-temperature limits for growth for a range of organisms.

Taxon	Temperature °C
Archaea	113
Cyanobacteria	75
Single-cell eukaryotes	60
Metazoa	50
Vascular plants	48

(principle sources: Pace, 2001; Rothschild and Mancinelli, 2001; Schwartzman, 1999). For lower temperature limits there is less variation. Some members of all major taxa can grow below 0°C. The lower limit is not clear but some microbes can grow at −20°C and the key to this is that the physical properties of ice allow solute diffusion at temperatures below 0°C (Pace, 2001).

see: Lenton, 1998; Lovelock, 2000a; Lenton and Wilkinson, 2003; Lovelock, 2003. For criticisms see Kirchner, 2002, 2003.) Whereas early versions of the Gaia hypothesis described 'atmospheric homeostasis by and for the biosphere' (Lovelock and Margulis, 1974) a more modern definition is 'that organisms and their material environment evolve as a single coupled system, from which emerges the sustained self-regulation of climate and chemistry at a habitable state for whatever is the current biota' (Lovelock, 2003, p. 769).

I will discuss Gaia theory in more detail in Chapters 6 and 11 although its influence is widespread throughout this book; at this point I will simply point out that it raises the interesting possibility that life may be involved in helping to maintain conditions suitable for its own existence on a planet such as the Earth. In this context it is worth asking, for any of my putative fundamental processes, what would be its long-term effect on the habitability of any planet on which it occurred? I have referred to this concept as the 'Gaian effect' of the process. Any process which tends to increase the survival of life on a planet is said to have a positive Gaian effect while one that decreases the chances of survival has a negative Gaian effect. This term has the advantage of brevity but it is not ideal since Gaia is usually defined as the whole system on a planet, thus using it to refer to part of the system is potentially confusing (Lenton and Wilkinson, 2003). However, I have stuck with the term (used in the original paper (Wilkinson, 2003) which was the forerunner to this book) as I consider the advantages of brevity outweigh the possibility for confusion and the term has the additional merit of making the origin of the idea, from Lovelock's Gaia theory, instantly apparent.

1.5. **Overview**

In this book I address the question 'what are the fundamental processes in ecology?' by considering the following thought experiment. *For any planet with carbon-based life, which persists over geological time scales, what is the minimum set of ecological processes that must be present*? I have identified seven putative fundamental processes which are described in the chapters constituting Part II of this book. I have not been overly concerned with a formal definition of 'process' in this context, other scientist would no doubt split (or lump) together some of my suggested 'fundamental' processes. Their main utility in this book is as a way of organizing ecology in a novel and hopefully thought-provoking way, where the emphasis is on placing ecology in a planetary— or Earth Systems—context. These processes (and their Gaian effects) are summarized in Table 1.2. Some processes, such as natural selection and

Table 1.2: Summary of the fundamental processes of ecology. The suggested Gaian effect (+, −, or ?) is given in parentheses.

Energy flow (required by the second law of thermodynamics), leading to energy consumption and waste product excretion. *Chapter 2.*

Multiple guilds: autotrophs, decomposers (and possibly parasites) (+). *Chapter 3.*

Tradeoffs leading to within-guild *biodiversity* (+). *Chapter 4.*

Ecological hypercycles (?; + if efficient nutrient cycling present). *Chapter 5.*

Merging of organismal and ecological physiology (+ or −). *Chapter 6.*

Photosynthesis? (not required but possibly likely in most biospheres). *Chapter 7.*

   Anoxygenic (+)
   Oxygenic (+ or −)

Carbon sequestration (+ or −). *Chapter 8.*

Nutrient cycling as an emergent property of the fundamental processes (+). *Chapter 9.*

---

competition, are characteristic of life in general rather than ecology in particular. As such they fall outside the main topic of this discussion, but will be considered in the context of several of the ecological processes in the following chapters. The final part of this book will consider the whole Earth system, in the context of these proposed fundamental processes.

Part II   The fundamental processes

We already have far more facts than we can handle. What we need most is to improve our ways of sorting and relating them—to work on concepts, to philosophize.

Midgley (1989, p. 53)

2. *Energy flow*

2.1. **The second law of thermodynamics**

The second law of thermodynamics is so important to many areas of science that the scientist and novelist C.P. Snow (1959) famously argued that knowledge of it was a crucial test of scientific literacy during his famous Rede lecture on 'The two cultures'. Snow thought of the Arts/Humanities as one culture and Science as the other, and suggested that few people in the Humanities would be able to pass his second-law-based Science literacy test. Although it seldom gets a mention in ecological discussions, the second law lies at the heart of fundamental ideas about energy flow in ecosystems.

At its simplest, the second law states that 'heat flows from a hotter to a colder body' so that entropy (Box 2.1) is greater (or at least not lower) after any given process has taken place (Penrose, 2004). At first sight organisms appear to violate the second law with their information-rich (low entropy) nature. Indeed organisms have been described as 'islands of order in an ocean of chaos' (Margulis and Sagan, 1995). The way in which organisms manage to achieve this is well known, but far from trivial, as it lies at the core of ecology. As is so often the case with overly familiar points, there is a real danger of ignoring them, indeed many ecology textbooks fail to emphasize the ideas I discuss in this chapter.

2.2. **Schrödinger, entropy, and free energy**

In 1944 the physicist Erwin Schrödinger published a short book entitled *What is life?* (I have referred to the slightly amended 1948 edition for reasons which will become apparent.) This book was influential in attracting many young physicists to biological problems in the mid-twentieth century (Perutz, 1987), and indeed Gould (1995) considered it amongst the 'most important books in 20th century biology', while Paul Davies (2005) has emphasized the irony that 'one of the most influential physics books of the twentieth century was actually about biology'.

Box 2.1:   **Entropy and information**

The principle sources for my approach in this box are Lovelock (2000a) and Penrose (2004).

Entropy (s) is the measure of the disorder of a system and is a notoriously difficult concept.

Formally:

$$s = K\ln \text{P}$$

where K = Boltzmann's constant (1.38×10^{-23} JK^{-1}), and P = a measure of disorder, effectively a measure of probability or the 'degree of surprise' (Aleksander, 2002).

Life is 'surprising' and as such has low probability (P) and hence low entropy (s). The idea of entropy is closely related to Claude Shannon's concept of information. Indeed information can be defined as $1/s$. Entropy can roughly be thought of as a measure of 'randomness' in a system whereas information can be thought of as a certain kind of lack of randomness.

A more formal definition of entropy requires the concepts of 'phase space' and 'coarse graining' (see Penrose 2004, chapter 27). However, for the arguments developed in this book all that is required is an understanding of the reciprocal relationship between entropy and information and a 'feel' for entropy as a measure of randomness (or negative entropy as a measure of surprise).

Although many of its more original ideas turned out to be wrong, it was influential in a way that defies simple summary (see discussions by Gould (1995) and Perutz (1987)).

In his book Schrödinger (1948, p. 71) famously described the process by which an organism survives as continually drawing negative entropy from the environment. In fact on thermodynamical considerations this is not strictly true, as pointed out by Franz Simon soon after the book was first published. In the 1948 edition of his book Schrödinger added a note to this effect, admitting that it might be better to consider organisms as drawing on free energy rather than negative entropy. These niceties of thermodynamic theory are not crucial for the concerns of this book. However, what Schrödinger was describing in thermodynamic terms is an important concept in ecology; namely that to survive all organisms must acquire energy from their environment and in so doing produce waste products which they release back into their surroundings. Indeed this is so fundamental that it could be considered *the* basic concept of ecology.

To a (very) naïve physicist, organisms appear to break the second law; however, they are not closed systems as they draw free energy from their surroundings, this is what enables them to give the *illusion* of cheating the second law. Just as a domestic refrigerator appears to violate the second law by moving heat from a cold to a hotter body, if the whole system is considered (in this example the refrigerator and its environment, which includes processes in the power station supplying the electricity) there is still a global increase in entropy although entropy decreases locally within the refrigerator. In a thermodynamic context the whole Earth system can be viewed in a similar way to an organism, for example Lenton (2004) described Gaia as 'a type of planetary-scale, open thermodynamic system, with abundant life supported by a flux of free energy from a nearby star'. In this respect it is sensible to consider Gaia as a superorganism, and the analogy is useful if one is considering energy (or entropy). However, clearly Gaia does not have 'genetics', so while a classical biochemist (interested in metabolism) could find merit in viewing the whole Earth System as a giant organism, a molecular biologist (interested in questions of information and genetics) would not. Like most metaphors, the idea that 'the Earth is alive' is useful in thinking about some questions but misleading for others. It is no doubt relevant that James Lovelock's background is as a chemist who worked for many years in medical research, prior to the great expansion of molecular biology, indeed he is happy to describe himself as 'an old fashioned organic chemist' (Lovelock, pers. comm.). However, many of the most important applied problems in the environmental sciences are essentially ones of the Earth's metabolism, for example changes in the carbon cycle. For these problems the metaphor of the living Earth has some merit.

Schrödinger's point about organisms and thermodynamics can be simply characterized as

$$\text{Energy} \rightarrow \text{organism} \rightarrow \text{waste product}$$

This is clearly very similar to the food chains common in introductory ecology texts, for example:

$$\text{Solar energy} \rightarrow \text{green plant} \rightarrow \text{herbivore} \rightarrow \text{carnivore}$$

However, in classical descriptions of food chains and food webs many of the waste products (CO_2, O_2, heat, etc.) are never shown, and many others (e.g. faeces or dead leaves) are only shown in special cases (Box 2.2). As no real system can be 100% efficient, any organism using free energy from its environment must be producing waste, which is then released back into the environment. As such there is nothing special in the fact that humans cause pollution (i.e. add waste to their environment), what is different about us is the size of the effect. Within my lifetime (I was born in 1963) the human population has more than doubled (Cohen, 1995), and in addition an increasingly affluent and industrial world is producing ever more waste. For example in 1990 there were approximately 478

Box 2.2:   **Illustrations of the partial nature of most food webs**

Here I briefly describe three food webs from different environments (terrestrial, freshwater, and estuarine) to illustrate the poor coverage of decomposers such as bacteria, protozoa, and fungi, in most food webs. Waste products are often ignored yet provide the free energy for crucial guilds of organisms (e.g. microbes involved in decomposition). For a recent, and rare, example of a detrital soil food web, see Mulder *et al.* (2005).

A terrestrial food web: Spitsbergen in the Arctic (Hodkinson and Coulson, 2004)

This food web concentrates on the multicellular invertebrates. The main decomposers are treated as a black box labelled 'soil microflora'. The exception to this are a group of nine taxa of testate amoebae identified by myself from soil samples collected by Hodkinson and Coulson. However, these data are the product of examining only seven soil samples (not 42 samples as incorrectly implied by Hodkinson *et al.* (2004)), while data on the macroscopic animals is based on substantially more work. The key point for the arguments in this book is that even in this highly detailed arctic food web the crucial microorganisms receive little detailed attention.

A fresh water food web: Tuesday Lake, Michigan USA (Cohen *et al.*, 2003)

This study focused on the lake's pelagic food web. It excluded the, probably non-trivial, nutrient flow into the pelagic web from the littoral zone and the terrestrial system surrounding the lake. More directly relevant to the topic of this box there are no microbial data (not even as a 'black box' as in the Spitsbergen example).

An estuary food web: Ythan Estuary, Scotland (Hall and Raffaelli, 1991)

This is an unusually complete food web, for example it contains 5,518 separate food chains. The web contains 92 species; however, microbes involved in decomposition within the system are relegated to a single black box labelled 'detritus'. Again this emphasizes the lack of detailed treatment of microbes even in unusually detailed food webs.

million cars in the world, by 2000 this had risen to approximately 560 million; emerging major economies such as China are predicted to greatly increase this number over the next few years (Myers and Kent, 2003). In the vast majority of cases of human pollution the problem is one of rate or magnitude, as most of our waste products also occurred on Earth prior to our evolution. There are some rare exceptions to this, such as chlorofluorocarbons (CFCs), which have no natural sources (Lovelock *et al.*, 1973), but normally with pollution we are altering the rate of production of a 'natural' product.

2.3. **Sources of free energy**

The source of energy we are most familiar with in Earth-based ecology is photo-synthesis, which is considered in Chapter 7. The other obvious source of energy for life as we know it on Earth are organic compounds, which would have been present on the early Earth, and may have been the first energy source for life on our planet. Organic compounds would have existed on Earth before life, indeed these chemicals are widely distributed in space; for example interplanetary dust particles currently add approximately 300 tonnes of organic matter to Earth every year (Gribbin, 2000). Studies of Comet Halley on its last close approach to the Earth in 1986 inferred the identity of a wide range of organic chemicals, ranging from simple compounds such as pentyne and hexyne to more complex ones such as iminopropene and xanthine (Kissel and Krueger, 1987). More exotic energy sources are also possible; for example, it has been suggested that an ecosystem on Europa (one of the moons of Jupiter) would be theoretically possible if based on organic molecules which could be produced by charged particles accelerated in Jupiter's magnetosphere (Chyba, 2000; Chyba and Phillips, 2001).

2.4. **Maximum entropy production and planetary ecology**

Entropy may also be important in the context of this book as a way of under-standing a planet's climate—clearly an important topic for ecology considered at a global scale. Classical thermodynamics tended to concentrate on equilibrium conditions, in much the same way that theoretical ecology focused on stable equilibriums in the mid-part of the twentieth century. However planets, like many ecological systems, are often non-equilibrium systems, this is especially true of planets likely to be of interest to a biologist (a 'heat death' planet is in a stable equilibrium condition which makes for simplified mathematics but little chance of interesting ecology). During the 1970s it was realized that heat transport on some planets may be organized to produce entropy at the maximum possible rate

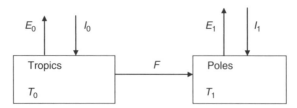

Fig. 2.1: Two box model of latitudinal heat flow on a planet such as Earth (after Lorenz, 2002 and Lorenz *et al.*, 2001). The two boxes are of equal surface area, *tropics* representing the planet's equatorial regions whereas *poles* represents the sum of the two polar regions. Each box has an associated absorbed solar flux (I_0 or I_1) and also an outgoing thermal flux (E_0 or E_1). The latitudinal heat transport from tropics to poles is 'F'. If neither tropics nor poles are showing a net change in temperature then: $I_0 - E_0 - F = 0$ and $I_1 - E_1 + F = 0$.

(Paltridge, 1975). This observation, which seemed to fit data for Earth, has been controversial—not least because the mechanism has been far from clear and these ideas have not been widely discussed outside a subset of climatologists and planetary scientists. As Lorenz (2002) has pointed out 'much of the literature discussing the principle is mired in forbidding algebra', which has reduced its wider impact. Having struggled with the primary literature on this topic as well as attending several related workshops in recent years, I can vouch for the truth of Lorenz's observation! However, there has been progress in producing a theoretical background for these observations which appears to put them on a firmer foundation (see Lorenz (2003) for a brief non-technical review).

The simple box model of a planet's latitudinal heat flow in Fig. 2.1 illustrates the main idea. In such a system the equation:

$$ds/dt = F/T_1 - F/T_0$$

is maximized (where s = entropy, see legend of Fig. 2.1 for the remaining notation). If the heat flow (F) was zero then the entropy production would be zero. If the heat flow was maximized the whole planet would be at the same temperature ($T_0 = T_1$) and again entropy would be zero. However, between these two extremes, entropy production is positive and there is a single maximum value. It appears that planets (e.g. Earth, Mars, and Saturn's moon Titan) operate in such a maximum entropy state (Lorenz *et al.*, 2001). One way of viewing such results is to suggest that the universal requirement for increased entropy (from the second law of thermodynamics) is responsible for the general circulation of the atmosphere on Earth and other similar planets (Ozawa and Ohumura, 1997).

In the context of this book the ideas of maximum entropy production (MEP) raise a number of interesting questions. On Earth, life is heavily involved in the dissipation of heat—for example, there is the role of vegetation in affecting the lands albedo or the role of life in the regulation of greenhouse gasses. If the MEP approach develops into a key way of thinking about planetary climates, then the presence of life on a planet will be a crucial part of the system. In addition, the MEP approach is in some ways similar to the Gaia hypothesis, at least in as much as suggesting that a planets system may be organized *as if* it was trying to maintain certain types of conditions over geological time. As such both ideas have attracted criticism because on a superficial reading they appear teleological. There are interesting initial attempts by Kleidon (2004) to integrate MEP and Gaia, based on thermodynamical considerations he argues that the presence of life introduces additional degrees of freedom into the system and as such, biotic activity should evolve towards MEP states.

2.5. **Overview**

Organisms maintain their highly ordered state by taking in free energy from their environment and (as use of this energy cannot be 100% efficient) releasing waste products back into the environment. This is so basic that it can be considered *the central concept of ecology*. The source of an organism's energy is a traditional topic of academic ecology; however, the nature and fate of the waste products is also of great importance and this has historically featured less prominently in ecological studies. The role of waste products, and their potential Gaian effect, will be considered in the next chapter. It is also possible that at a global scale ideas of maximum entropy production may be very important in understanding a planet's climate, because of its role in heat dissipation, life can have a major role in such systems. Most of these MEP-based ideas are very new and it is as yet unclear how important they are in understanding planetary scale processes.

3. *Multiple guilds*

3.1. **The importance of waste**

In the previous chapter I argued that the key idea at the heart of ecology is that organisms must use their environment both as a source of energy and as a dumping ground for their waste products. Also that any process which organisms use to extract this energy cannot be 100% efficient. I described how these ecological ideas are linked to crucial concepts from the physical sciences such as the second law of thermodynamics and entropy. These arguments can be summarized as:

$$\text{Energy} \rightarrow \boxed{\begin{array}{l} \text{Organism} \\ \text{some energy used} \end{array}} \rightarrow \text{Energy (waste)}$$

As such, there is still potentially free energy available in the waste products of organisms: as illustrated by the smaller font size used for the energy in waste products in this schematic representation. These waste products can potentially have important effects on other species, as illustrated by the problems caused by 'pollution'; that is, human waste products.

One approach to trying to visualize the quantity and importance of waste products produced by the Earth's biota is to think about the temperate deciduous forests that dominated much of Europe, eastern North America, China, and Japan, before being reduced by human actions. Every year, at the end of summer, the leaves on the trees are transformed from the greens of chlorophyll to the reds and yellows of anthocyanins and carotenoids, before falling to the forest floor. These leaves, along with the other plant and animal debris produced over the year amount to approximately 5 tonnes of biological waste per hectare per year. The figure for highly productive tropical forests can be an order of magnitude higher, at approximately 60 tonnes per hectare (Spooner and Roberts, 2005). It is common observation that something must happen to get rid of this waste, since when we enter a forest we are not up to our necks in leaf litter!

The main strategy of this book is to use astrobiological thought experiments to give a wider perspective to thinking about Earth-bound ecology. As such it is worth considering the simplest possible biosphere, a hypothetical planet inhabited by only a single species; over time this species would convert its resources into both copies of itself and waste products. As these individuals died they would also be added to the waste product pool. Eventually such an ecological system would run down as the species ran out of crucial resources, which had become locked up in waste products. As Tyler Volk (1998, p. 52) has written, 'A monoculture planet is therefore a thought experiment with no place in reality.' I will return to thinking about this hypothetical planet in more detail in Chapter 9.

Clearly in any successful ecological system it is necessary for material in waste products to be recycled to make them available for reuse (Fig. 3.1). Any ecological system that produced an intractable waste product which couldn't be broken down would, on 'geological' time scales, be in trouble. A classic example of a biochemical product which is unusually difficult to break down is lignin (Box 3.1). Shortage of lignin-degrading guilds in the later Palaeozoic may explain the large coal deposits formed at this time. This build up of plant material in sediments would have been exacerbated by the fact that the large plants of

Fig. 3.1: This beach on the Spanish island of Mallorca shows what happens when a substance which is hard for organisms to break down (plastic) is added to a system over time. Plastic waste can now be found in quantity even on very remote beaches around the world (Barnes, 2002), as it is constantly being added to the marine system but only slowly broken down.

Box 3.1:   **Lignin and why its so hard to biodegrade**

Lignins are phenylpropanoids and are the second most abundant polymer in most plants (after cellulose). They provide strengthening for cell walls and xylem vessels. They are also very important compounds in the Earth system, it is estimated that phenylpropanoids contain 30% of all organic carbon in the biosphere, most of this in the form of lignin (Davies, 2004). The burial of such compounds (e.g. in the Carboniferous coal deposits) has implications for the carbon and oxygen cycles which are discussed in Chapters 7 and 8.

There are three main reasons why lignin is difficult for organisms to break down (Robinson, 1990).

1.   Molecules of lignin are insoluble in water and are too heterogeneous to be easily disassembled by specific enzymes.

2.   Many lignin degradation products are themselves difficult to break down and/or are toxic to many organisms.

3.   The C : N ratios in lignins are so high that few organisms could live on lignin even if they could break it down into usable compounds. Some of the fungi which can break down lignin get round this problem by trapping and digesting nematode worms using special hyphae, so partly subsidizing their nitrogen intake. Examples of fungi which use this strategy include the Oyster Cap fungi *Pleurotus spp.*, which specialize in growing on wood (Spooner and Roberts, 2005).

the time (e.g. Club Mosses and Ferns) used more lignin in their structure than later woody plants (Robinson, 1990). The key members of the modern lignin degrading guild are fungi, particularly Basidiomyetes, many of which produce the familiar 'mushroom' fruiting bodies. These 'white rot' fungi break down both lignin and cellulose, their food source is the cellulose, however, they need to break down the lignin enzymatically to get at it (Fig. 3.2). This process leaves the remaining wood looking 'stringy, bleached and soft' (Spooner and Roberts, 2005). Had lignin consuming organisms not evolved this could have proved problematic for abundant life on Earth, with increasing amounts of resources entombed in organic deposits and unavailable to life. While simple microbial systems may be able to continue in such conditions, it is difficult to see how an active and complex biosphere could be supported with such a lack of recycling.

Fig. 3.2: The Stinkhorn *Phallus impudicus* is a common species of 'white rot' fungus in Britain, mainly colonizing dead wood after it has fallen to the forest floor (Spooner and Roberts, 2005). Such lignin degrading organisms are crucial in the breakdown of woody plant remains. In addition to its wood-rotting ability the fruiting body of the fungus is famed for its phallic shape, celebrated in its scientific name, and its obnoxious smell which attracts flies and slugs to aid its spore dispersal (Ramsbottom, 1953). The example in the photograph has just finished spore dispersal, the blackish liquefied spore mass having been largely removed from its head by visiting invertebrates.

3.2. **The requirement for multiple guilds**

Biological waste is also a potential resource for other organisms. For example the CO_2 we breath out is a raw material for photosynthesis by plants, just as the oxygen produced as a by-product of photosynthesis is crucial to aerobic organisms. The bodies of other organisms (alive or dead) are clearly composed of materials needed by life. In addition the fact that no energy extraction process is 100% efficient means that waste products of the organisms can provide a potential source of resources for other life forms, which will often quickly recycle these resources (Fig. 3.3).

This last point is well illustrated by the effects of various veterinary antiparasitic drugs on the degradation of animal dung. One of the most well known of these drugs is ivermectin, which is used to control both endoparasites (e.g. lung worm) and ectoparasites (e.g. lice) in domestic animals such as cows, horses, and sheep (Floate *et al.*, 2005). Since the 1960s there has been a concern that

Fig. 3.3: *Panaeolus semiovatus* fruiting bodies on horse dung on Dartmoor in south west England, illustrating the important ecological idea that one organism's waste is another's energy source. This species appears to specialize on horse dung (Kibby, 1979).

dung containing these chemicals may break down more slowly, negatively affecting both plant growth on the affected pastures and the invertebrate species which specialize on feeding on animal dung (Table 3.1). Although the effect has not been found in all studies, many researchers have found that dung containing chemicals such as ivermectin decomposes more slowly (Floate *et al.*, 2005), thus illustrating the way in which one species' waste product can be a resource for another species.

Even though it seems clear that these chemicals in dung can reduce the activity of invertebrates such as beetles and flies, surprisingly little is known about the potential effects on microbes and other organisms. Experiments by

Table 3.1: Summary of the total numbers of invertebrates recorded in control and ivermectin contaminated experimental cow pats (dung) on a farm near Bristol in England.

Cow pat type	Number of samples	Coleoptera	Diptera
Control	16	1,229	286
Ivermectin	17	29	40

The numbers of coleopteran (beetles) is the sum of all individual larvae, pupae and adults recorded while the figures for diptera (flies) is for larvae and pupae. The experiments were sampled over a time span of 20–100 days. Data summarized from Wall and Strong (1987) who provide a more detailed tabulation of these data.

Sommer and Bibby (2002) showed that ivermectin still had an effect on dung degradation when larger invertebrates were excluded, but their experiment did not exclude animals which could get into 1 mm mesh bags. Therefore, ivermectin could have been affecting very small animals and/or microbes. If, as most workers in the area seem to assume, the main effect of these chemical on dung degradation is through its effects on invertebrates rather than microbes then this system provides a good example of animals acting as 'catalysts' in decomposition, speeding up the process over the rate when just relying on microbes (see Section 1.1). However, it is probably premature to rule out any effects of ivermectin on microbes without further study and the same is true of its possible effects on other organisms. For example, Porley and Hodgetts (2005) speculated that ivermectin may be involved in the decline of *Splachnum ampullaceum* in Britain, this moss grows on the fresh dung of grazing animals, and as with the possible effects on microbes there is a lack of data to evaluate its effects on mosses.

Whatever the role of microbes in the ivermectin example a catalytic role for animals certainly appears to be the case with some more natural dung-based systems. For example, in eastern Kenya up to 48,000 beetles have been observed visiting a 3 kg pile of Elephant *Loxodonta africana* dung during a 2-hour period. During this time the entire pile of dung was eaten *in situ* or carried off to be buried by the beetles (Coe, 1987). This behaviour greatly speeds up the rate of microbial breakdown of the dung, although it should be noted that much of the consumption of the dung by beetles is facilitated by mutualistic symbiotic microorganisms living within the insects, so that a microbiologist would have some justification in viewing dung beetles as animated microbial systems!

These examples suggest that an ecological system needs organisms from at least two major guilds, autotrophs and decomposers (which release material from waste products, including dead autotrophs). As such any ecological system which has survived for a 'geological' period of time on a planet must consist of at least these two types of organisms. But what of other major guilds, such as predators and parasites?

3.3. **Parasites and predators**

Predators, and their prey, often feature very prominently in ecological studies. They are regularly the stars of television nature documentaries, which are (unfortunately) probably most people's main source of ecological knowledge. One of the most well-known serious 'popular' books on academic ecology is Paul Colinvaux's *Why big fierce animals are rare* (Colinvaux, 1980), which gives large predators pride of place in the title. Similarly, the most comprehensive single volume text book in ecology (Begon *et al.*, 2006), devotes 59 pages to predators but only 20 pages to decomposers. Yet predation, as a way of life, is not fundamental to ecology (unlike decomposers). Indeed, *if* there is life on other planets my expectation is that predation *sensu stricta* will only occur on a minority of them, since my guess is that most of them will not have progressed beyond bacteria-like organisms. This guess is based on the assumption that simpler types of life are likely to be more common than complex types, and the probability that eukaryotes took a considerable time to evolve on Earth. However, these assumptions could easily be undermined by a more detailed knowledge of the history of life on Earth or the discovery of life in other parts of the universe.

A reasonable definition of predation is that it is the consumption of—often whole—organisms by carnivores leading to the immediate death of the prey (Krebs, 2001). In the history of life on Earth this probably required the evolution of phagocytosis, which is restricted to animals and protists (Margulis and Schwartz, 1998). As such, predation *sensu stricta*, where the predator engulfs the prey, would not have been present until the evolution of the eukaryotes. Modern protozoa use phagocytosis to prey on a wide variety of organisms including bacteria, other protozoa, algae, fungal hyphae and spores and small metazoans (e.g. Gilbert *et al.* (2000) for testate amoebae: Foissner (1998) for soil living ciliates). The date of this major stage in evolution is not known with much certainty, however currently the earliest eukaryote fossils are often claimed to be between 1.7 and 1.9 billion years old, although these dates are controversial. As well as claimed traditional fossils there are sterols, chemical 'fossils' (interpreted as the diagenesis products of cholesterol) in rocks of 2.7 billion which may suggest the presence of eukaryotes (Carroll, 2001; Nisbet and Sleep, 2001; Margulis and Dolan, 2002). However it is now clear that a prokaryote origin for these sterols cannot be ruled out, in addition the evidence from molecular phylogenies for the origin of eukaryotes is also 'deeply controversial' (Nisbet and Fowler, 2003). As such the age of origin of eukaryotes has become as unclear as the date of the earliest life discussed in Chapter 1.

One way to view phagocytosis is that it helps side step a potential problem with bacterial extra cellular digestion. Prokaryotes are unable to take food particles directly into their cells, so they have to release enzymes into their environment to break their food down into chemicals which can enter their cells. This raises the potential problem of cheating bacteria which rely on the extra cellular enzymes of others rather than producing these potentially costly chemicals themselves. In theory there are at least two ways bacteria can reduce this problem. First, it would seem logical that the less these enzymes diffuse into the wider environment, the harder it is for the cheaters to do well. This commonsense interpretation is consistent with the limited formal modelling of the problem by theoreticians (Allison, 2005; Sherratt *et al.*, 2006). Second, while we need to know much more about the mechanisms of extra cellular digestion, it appears that many bacteria have adaptations to keep these chemicals in the immediate vicinity of their cell surface (Fenchel *et al.*, 1998). As such it may be an oversimplification to think of most of these enzymes diffusing through the microbe's environment. However, it seems an obvious speculation to suggest that the cheat problem with extra cellular enzymes may have been one of the selection pressures leading to the evolution of phagocytosis. From an ecological perspective this produced the first true predation on Earth.

Once autotrophs and decomposers exist then it is possible that parasites, capable of using resources from other living organisms while they are still alive, may have been quick to evolve. Observation of organisms on Earth shows that the lack of phagocytosis is no barrier to parasitism (e.g. parasitic viruses, bacteria, fungi, and plants; Fig. 3.4). If Earth is at all typical then natural selection seems very powerful in targeting available resource pools and so parasitism could be common to all biospheres (see Box 3.2 for some complications with definitions). With Earth the only known example of a planet with life, how typical it is presents a big problem! However, selection is also powerful in targeting resources in artificial life models (e.g. Downing and Zvirinsky, 1999) suggesting that in this respect the Earth may well be typical.

Fig. 3.4: Broomrapes are parasitic plants which do not photosynthesize but obtain their resources from the roots of other plants. This is Slender Broomrape *Orobanche gracilis* a parasite of legumes, growing in the mountains of northern Italy.

Box 3.2: **Parasites and predators, problems of definition**

Like most ecological definitions 'parasite' and 'predator' are, at least in part, artificial classifications of convenience (Fig. 3.5). While a protozoan engulfing a bacteria or a big cat eating a gazelle are clearly cases of predation, some examples are more difficult to categorize. To illustrate the problem two such cases are highlighted in this box.

Predation without engulfing the prey?

Both carnivorous plants and nematode trapping fungi (see Box 3.1) catch and kill prey but extract the nutrients by extra-cellular digestion. As the whole organism is killed and consumed this would normally be described as predation rather than parasitism. Indeed, Darwin (1875) wrote of insectivorous plants that they 'may be said to feed like an animal'. However, the vast majority of examples from natural history which are usually classed as predation involve the prey being engulfed.

Box 3.2:   (*Continued*)

Parasitism with (partial) engulfing of the prey

Many of the familiar multicellular parasites (especially ectoparasites such as fleas and lice) engulf a small part of their host without killing it. Such organisms are always traditionally classed as parasites. However, many plant-eating animals also only eat a small portion of the plant (without killing it). Arguably this is very similar to lice and fleas; however, caterpillars and deer are seldom described as parasites. One difference here is the nature of the prey. Plants are usually modular organisms, and loss of some modules (leaves or braches) will not kill them. This is much less the case with animals, although many invertebrates can lose legs, or some lizards their tails, without death ensuing.

(a)

Fig. 3.5:   (a) A carnivorous plant, the Pitcher Plant *Sarracenia purpurea* growing on a 'floating bog' in Ontario, Canada. (b) Parasite or predator? Caterpillars of the Cinnabar Moth *Tyria jacobaeae* feeding on Ragwort *Senecio jacobaea* in Lincolnshire, eastern England.

Box 3.2:   *(Continued)*

(b)

Fig. 3.5:   *(Continued)*

3.4. Parasites introduce a potentially important mechanism for density-dependent regulation

It is well known that parasites, especially some microparasites such as viruses and bacteria, can greatly effect the host's population size. For example, the Black Death in the fourteenth century killed a substantial proportion of the human population of Europe; estimates of between 25% and 50% mortality are commonly cited (Anderson and May, 1991; Cohen, 1995), although based on limited data. In some unfortunate places the mortality was much higher, one example was the Abbey of Meaux in northern England where only 10 monks, out of an original compliment of 42, survived the first wave of the Black Death (Schama, 2000). More recently a large number of people were killed in the influenza pandemic of 1918–1920, historically the estimates for the global death toll have been around the 20 million mark (Anderson and May, 1991; Porter, 1997), however, more recent studies of contemporary records suggest a figure of 50 million or more (Johnson and Mueller, 2002). As I wrote the first draft of this chapter (June 2005) there was a growing concern about the possible effects of the next influenza pandemic in a crowded world of over 6 billion people (Osterholm, 2005), indeed it is very plausible that some readers of this book could die in such a future pandemic (emeritus professors should probably worry more than university students, as in many influenza outbreaks the elderly are particularly at risk, although the 1918–1920 pandemic apparently did not follow this pattern). Crowded populations are more susceptible to such virulent microparasites, indeed this is why many authors assume that these microparasites were less important in human history before the development of agriculture and large settled populations (Anderson and May, 1991; Cohen, 1995; Diamond, 1997).

These examples, of the effects of microparasites on populations of our own species, illustrate the potential for parasites to reduce population sizes, and potentially to act in a density-dependent manner. That is, help to regulate population size by increasing mortality at high population density while being a very minor threat—so increasing an individual's probability of survival—at low population densities. Because of this it is very difficult for such a parasite to drive a host to extinction, at least in relatively simple mathematical models (Anderson and May, 1991). One exception to this simple theoretical story are parasites which show frequency dependent transmission rather than density-dependent transmission. Obvious example are sexually transmitted diseases which can certainly drive hosts to extinction in mathematical models. Another exception is where the parasite has a wide range of hosts or even can survive by non-parasitic means, so density reduction of one host does not reduce the chances of infection. An example of this is the fungus *Batrachochytrium dendrobatidis* which has apparently caused extinctions of amphibian species in South America (where the fungus is introduced), in this case climate change appears to be exacerbating the problem (Pounds *et al.*, 2006). However, in a review of the empirical literature, de Castro and Bolker (2005), found very few well-documented cases of disease-induced extinction, suggesting it is probably rare in nature (exceptions may be when diseases affect populations which are already in trouble for other reasons).

The basic idea of density-dependent regulation has a history longer than that of academic ecology. For example, it is implicit in the ideas of T.R. Malthus on human population at the end of the eighteenth century. His conclusions—about population crashes being likely at high numbers—are perhaps best known to biologists through their claimed influence on the thinking of Charles Darwin (Mayr, 1982). One of the great advantages of mathematics over verbal reasoning is that it is usually much less ambiguous, as such the logistic equa-tion—which features in all ecology textbooks—makes clear the assumption that population growth declines as population size increases (although it does not explain why). This equation was first used by Pierre-Francois Verhulst in the first part of the nineteenth century, although it was more famously rediscovered by Raymond Pearl and Lowell J. Read in the early twentieth century (Kingsland, 1995). Although parasites are not required for density dependent regulation, they can certainly cause it, through affecting fecundity and/or sur-vival. This regulation can be caused by marcoparasites such as fleas, mites, and nematodes as well as the microparasites described above (Tompkins and Begon, 1999). Shortage of resources and/or increased 'pollution' from waste products when organisms are common can also have density-dependent effects and would even operate on a hypothetical planet populated by only one species, and so parasite free.

Density dependence is important in the context of this book's main thought experiment as it has the potential to have a positive Gaian effect. As such the presence of parasites on a planet may also tend to increase the probability of life surviving there for longer. This point is illustrated for single species populations by a simple Monte Carlo simulation model which shows the effects of adding density dependence on the probability of a population's survival. The model was run using VORTEX (Lacy, 2000: Miller and Lacy, 2003), one of the commonest population viability models used in conservation biology. The results (Fig. 3.6) show that adding density dependence to the model increased the life expectancy of populations of a hypothetical species, modelled to have a similar biology to the leopard (see legend of Fig. 3.6 for more details).

A real world example from conservation biology is an island population of a sub species of Silvereyes *Zosterops lateralis chlorocephalus* off the Australian coast. This is a small passerine bird species and a population viability model of this island population was described by Brook and Kikkawa (1998), using an ear-lier version of VORTEX which did not have the facility to explicitly model dens-ity dependence. Their model suggested that this population was at some risk of extinction over a 100 year time period. Subsequently, McCallum *et al.* (2000) used an analysis of field data to demonstrate that this population exhibited density-dependent regulation, with a negative relationship between population size and population growth rate. They used their results to construct a simple density-dependent model of the population which predicted a much lower probability of extinction than did the original VORTEX model, which ignored density dependence. So, as in my simple model used to generate Fig. 3.6, adding density dependence increased the predicted probability of survival of this population. Analogy suggests that density-dependent processes will also be important at a planetary scale, so potentially increasing the time span of life on a planet. One

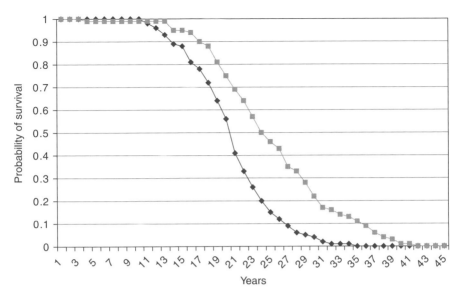

Fig. 3.6: VORTEX model of the effect of density dependence on the probability of survival of a population. The modelled organism is based on the leopard *Panthera pardus*. Model variables follow Wilkinson and O'Regan (2003) except that a carrying capacity (K) of 200 was used (with a 50 : 50 sex ratio). In all runs of the model the population started out at K and the estimate of probability of the population's survival was based on 500 iterations. This model was run using VORTEX version 9.21 (Millar and Lacy, 2003). For the population marked by diamonds (♦) only a ceiling carrying capacity was used, while for the population marked by squares (■) full density dependence was modelled (but without an Allee effect). When an Allee effect was included (in a series of numerical experiments using a range of plausible values) the result was always intermediate between the results for ceiling carrying capacity and full density dependence without an Allee effect (to keep the graph simple these results are not shown). While there are many problems with using VORTEX to estimate the extinction probabilities of real populations (e.g. Harcourt, 1995; Coulson *et al.*, 2001), here it is used in a qualitative manner to explore and illustrate the effects of altering model assumptions (see O'Regan *et al.* (2002) for an example of a study with a similar philosophy). The key point is that in this, admittedly simple, model the effect of adding density dependence is to increase the probability of population survival.

can imagine a microbial species becoming very common and potentially over exploiting its environment, only to be controlled by being targeted by a parasite. Many years ago J.B.S. Haldane (1932) argued that parasites played an important role in the evolution of species, they may be similarly important in the evolution of planets with life.

The discussion so far is an oversimplification, as density dependence does sometimes have the potential to *increase* the risk of population extinction. An obvious example being cases where increasing group size improves an individuals chance of survival, such as some group living animals where a larger group may have greater chance of spotting predators (Allee *et al.*, 1949). This process is usually referred to as the Allee effect, after Warder Clyde Allee who developed these ideas during the 1930s and 1940s. Since then many ecologists have assumed that this idea was of limited importance, however, during the 1990s there was a growing realization that the Allee effect may be important to the population dynamics of many organisms (Courchamp *et al.*, 1999; Stephens and Sutherland, 1999; Bessa-Gomes *et al.*, 2004). Much of the recent interest has come from behavioural ecology, for example studies of African Wild Dogs *Lycaon pictus* suggest that because they are cooperative breeders they may exhibit a critical group size below which extinction becomes very likely (Courchamp *et al.*, 1999). Unfortunately, due to our effects on the environment, many species of larger animals now live in small populations and this is probably one of the main reasons for the recent growth in interest in the Allee effect.

Even if life is widespread on other planets, it is quite possible that species of the behavioural complexity of Wild Dogs may turn out to be very rare. However, the idea of the Allee effect also applies to organisms, such as bacteria, which are more central to the key themes of this book. For example, it has become increasingly obvious that many microbes live in groups forming the so-called biofilms (Atlas and Bartha, 1998). These microbes appear to benefit from growing in such aggregations, for example it can increase their resistance to antibiotics, both in medical situations and presumably also in the wild where other microbes may produce antibiotics as part of inter-specific competition (Hoffman *et al.*, 2005). As such these microbes may show inverse density dependence—the Allee effect—at small population sizes because of a reduced capability to resist antibiotics. Some of the earliest uncontroversial evidence for life on Earth are the fossils of microbial colonies known as stromatolites, it is easy to imagine that Allee effects could have been important in the survival of these colonies.

3.5. **Other effects of parasites**

As well as their effects on population dynamics—which are the most important for this book's principle thought experiment—parasites are also important in an evolutionary context. Once parasites exist it is clearly in the interests of hosts to evolve strategies to protect themselves. For example, the pupa of some species of ladybird—ladybug in north America—beetles (Coccinellidae) have evolved a defence against some species of parasitic flies and wasps which lay their eggs in the pupa (technically such insects are often called parasitoids). If the pupa detects the touch of such an insect, it can flick its anterior end quickly up and down, so making it more difficult for the parasite to lay its egg (Majerus, 1994). Ladybird pupae can sometimes be persuaded to exhibit this behaviour if experimentally touched with a blade of grass, mimicking the touch of a parasitoid.

More importantly, a good case has been made for parasites being involved in the evolution of sex, or at least the maintenance of sexual reproduction in populations once it has evolved. These ideas are particularly associated with W.D. Hamilton (see Hamilton (2001) for collected relevant papers and Wilkinson (2000) for a brief review of the history of these ideas). Hamilton argued that, by repeatedly generating rare genotypes, sex made it harder for parasites to evolve adaptations to their host's defences—his argument was based on the essentially frequency-dependent ideas of J.B.S. Haldane who pointed out that parasites will mostly target common genotypes. These ideas have been very influential; however, a full explanation of the evolution and maintenance of sex probably requires additional ideas alongside Hamilton's parasite-based theory (West *et al.*, 1999), including ideas from genetics which are not directly relevant to this book's ecological thought experiment. However, if Hamilton was at least partly correct then it raises the possibility that we would expect to see some form of sex on most planets with life.

3.6. **Overview**

The problems faced by a hypothetical planet with only one species strongly suggest that any functioning ecological system must have organisms from at least two major ecological guilds, namely autotrophs and decomposers. Although conventional predators do not seem to be crucial to planetary ecologies it is likely that parasites will quickly evolve and through density-dependent processes help to regulate population sizes. Density dependence—through lack of resources, build up of waste products or parasites—may be crucial in preventing the runaway population growth of a species, leading to it monopolizing a planet's ecology. Although density-independent processes (be they a cold winter on a local scale, or the impact of a large meteorite at the planetary scale) can greatly affect abundance they cannot provide regulation, this requires the 'thermostat' like behaviour of density dependence. As such, both multiple guilds and the presence of parasites are likely to have positive Gaian effects in most biospheres. Indeed, parasites are likely to be an important selection pressure on most planets with life. This argument for multiple guilds has been based primarily on guilds defined by the way their members access carbon (as autotrophs or various guilds of consumers); however, similar arguments could be made for other types of biochemical guilds (sensu Volk, 1998), for example, the biochemical guilds of the nitrogen cycle such as nitrogen fixers and denitrifiers.

4. *Tradeoffs and biodiversity*

4.1. **The problem of biodiversity**

The diversity of nature, exemplified by Darwin's entangled bank, is so familiar from everyday experience that it is easy to accept it without much critical thought. We are so used to a multitude of species, and higher taxa, that it is difficult to appreciate the fundamental theoretical problems this diversity presents.

Consider these data from a 15×15 cm quadrat I recorded at an altitude of 1,890 m in the Dolomites (northern Italy). Within this very small area of ground in a coniferous forest I found mosses, sedges, grasses, and broad-leaved plants from three different families (Asteraceae, Ranunculaceae, and Ericaceae) along with a small pine seedling. With the exception of the seedling this quadrat was quite typical of the forest floor at this location. However, green plants can be considered as just ways for life to access solar energy, so why does it take so many different types of 'solar panels' to do this job on 225 cm^2 of ground in a forest on an Italian mountainside? An even smaller scale example comes from the bark and wood chip mulch used on some of the soils on the science campus of my University in Liverpool. Using a microscope I counted 100 individual testate amoebae (Fig. 4.1) from a small sample of this mulch and found they belonged to at least 14 species from 10 different genera (since I made no attempt to split several difficult taxa which appear nearly identical under conventional light microscopy, the real species richness was probably slightly greater). Why is more than one species of protozoan needed to do the job of eating bacteria and other small organic particles in this apparently simple habitat?

These two examples illustrate biodiversity at the small scale; plants on 225 cm^2 of ground or protozoa in less than a gram of bark chippings. However, the problem is also apparent at the planetary scale. Robert May (1988), in a famous paper in *Science*, drew attention to the fact that we have very little idea as to how many species currently inhabit the Earth. One result that most authors agree on is that insects appear to be especially rich in species, for example in the

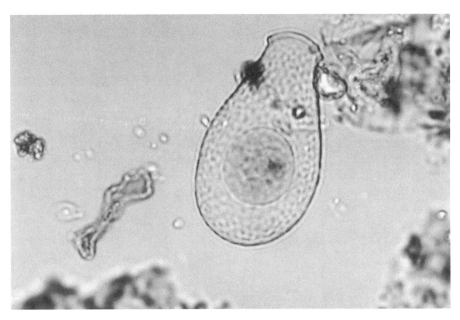

Fig. 4.1: Shell of the testate amoebae *Nebela collaris sensu lato* one of the species found in the wood chip mulch on my University campus in Liverpool, UK. This species is typically between 100 and 180 μm long and feeds mainly on immobile microorganisms (such as micro algae and fungal mycelia) along with senescent and dead microbes (Gilbert *et al.,* 2003). The encysted organism can be seen within the shell, when active its pseudopodia emerge from the 'mouth' of the shell to collect food particles.

early 1980s Erwin (1982) estimated that there were at least 30 million insect species, a conclusion he reached by extrapolation from the ratio of described to undescribed species in a study of beetle species richness in trees in Panama. Using a similar extrapolation approach, but based on data on hemipteran bugs in Indonesia, Hodkinson and Casson (1991) estimated a lower figure of approximately 2 million, although some other estimates have reached 80 million (reviewed by Stork and Gaston, 1990). The actual numbers produced by these estimates are best treated with a high degree of scepticism, however, they do make two important points.

1. The range of estimated numbers of insect species (from 2 to 80 million) strikingly illustrates our current lack of quantitative knowledge of biodiversity. Since the 1980s there has been a general consensus that the highest values are probably overestimates (Purvis and Hector, 2000; Novotny and Basset, 2005), although it is still impossible to answer the question 'how many insect species?' to the nearest 10 million with any great confidence!

2. Even the lowest figures, such as Hodkinson's and Casson's (1991) estimate of 1.84–2.57 million species are still surprisingly large if one considers the obvious functional questions, 'what do all these species do?, and why are there so many?'

One of the reasons for concentrating on insects in the above analyses is that they form a significant proportion of all described species (Wilson, 1992) so if we had a good estimate for insects we could make an educated guess at the total species richness of all organisms on Earth. However, this assumes that the relatively low numbers of described species of microorganisms are a real measure of their diversity and not an artefact of them being small and difficult to study; a crucial point appreciated by May (1988) in his *Science* paper. Currently, the extent of geographical isolation, and hence presumably species richness, in microorganisms is very controversial, this is not helped by the difficulty in defining 'a species' for many microbial groups. Some high-profile studies have been claiming that almost all free living protozoa show a complete lack of geographical isolation (see Finlay 2002 or Finlay *et al.*, 2004 for reviews), however, other authors suggest that although protozoa are more likely to be cosmopolitan than larger organisms there is still evidence that many species have limited ranges (e.g. Foissner, 1999; Hillebrand *et al.*, 2001; Wilkinson, 2001a). By analogy what applies to free living protozoa may also apply to other groups of microbes such as bacteria (Finlay and Clarke, 1999), so the outcome of this debate could greatly affect the estimated number of species on Earth; although there is always the possibility that different microbial groups may differ in their behaviour. If free living microbes are very species-rich then calculations that assume insects make up approximately half of all species will be spectacularly wrong (this assumption would also be wrong if it turns out that the average insect has several species-specific parasites, although in this case insect species richness could still be used as a guide to total species richness). This debate also has implications for conservation biology, if some microbes have limited ranges then our actions could cause their extinction much more easily than if their ranges are global with the species occurring anywhere in the world where the correct habitat is found (Finlay *et al.*, 1997).

The previous chapter discussed reasons why any planetary ecology should consist of two or more principal guilds, such as photosynthetic autotrophs or decomposers, and argued that multiple guilds are so fundamental that they would be found on any planet with life. However, those arguments give us no reason to expect that these guilds should be subdivided into large numbers of different types (species), which is what we observe on Earth. In this chapter, I consider one of the fundamental processes which can cause such biodiversity—the idea of tradeoffs. Without tradeoffs the role of species-rich guilds could be filled by single taxa, such as a 'perfect' photosynthetic plant. Such 'Darwinian Demons' would dominate a planet leaving no room for speciation. I argue that these trade-offs would tend to cause within-guild biodiversity on any planet with life. I also consider the Gaian effect of this resulting biodiversity, a topic of more than academic interest as we are currently in the process of greatly reducing biodiversity on Earth through our actions.

4.2. Tradeoffs illustrated by human sporting performance

Consider the following description of a group of sportsmen; 'Of the eight, five are tall and slim, long-limbed and lightly-muscled; the other three are muscular and compact' (Perrin, 1990). This short 20-word description is enough to eliminate a number of possible sports, they clearly do not appear to be jockeys or sumo wrestlers, while the 'compact' individuals make it unlikely it is a description of a group of basket ball players. That five are 'long-limbed and lightly-muscled' tends to rule out weight lifters, rugby players, or shot-put specialists. This brief fragment of prose is actually part of a description of a group of high-standard rock climbers at a cliff in the English Peak District in the 1980s and illustrates the point that different types of physique tend to be associated with success in different sports. To the uninitiated it may seem obvious that in rock climbing the more muscle you have the better, especially on overhanging rock. However, the main work done by a rock climber is to move their body weight up the rock face against gravity, since muscle is a heavy tissue what really matters is the climbers' power to weight ratio, this is a tradeoff between amount of muscle and body weight. So a good rock climber's body is more likely to look like that of a gymnast or a professional dancer than a weight lifter or body builder.

These ideas have been developed with more quantitative rigour in an analysis of the performance of decathletes by Raoul Van Damme and colleagues. The decathlon comprises 10 different track-and-field events and Van Damme *et al.* (2002) constructed a data set of the performance of 600 'world-class decathletes' from information available on the Internet. The effects of tradeoffs were apparent in the comparison of performance in some of the events. For example, there was a negative correlation between performance in the 100 m and the 1,500 m, also between performance in the shot-put and 1,500 m. Overall there was a highly significant negative correlation between overall performance across the 10 events and excellence in a particular specialist discipline. The sort of people who tend to win decathlons do well in all the disciplines but may lose to a specialist in any particular event. Van Damme *et al.* (2002) explicitly used their analysis as a novel way of quantifying the effects of specialization and tradeoffs in organisms. They concluded that 'in an environment in which the selection criterion is combined high performance across multiple tasks, increased performance in one function may impede performance in others'.

4.3. **Tradeoffs in ecology**

In an ecological context a good definition of tradeoff is as 'an evolutionary dilemma whereby genetic change conferring increased fitness in one circumstance inescapably involves sacrifice of fitness in another' (Grime, 2001, p. 10). This process appears to have grown in importance in ecological theory in the last 15 years as illustrated by an increase in the use of the term in the ecological literature since 1990 (Fig. 4.2). While Grime (2001) gives the formal definition of tradeoff cited above in his recent edition of *Plant strategies, vegetation processes and ecosystem properties*, the first edition of this book, published over 25 years ago (Grime, 1979), has no formal discussion of tradeoff. Although the idea of tradeoff was implicit in Grime's discussions in the first edition he does not emphasize the concept or list the term in his index, illustrating the recent rise in interest in this idea.

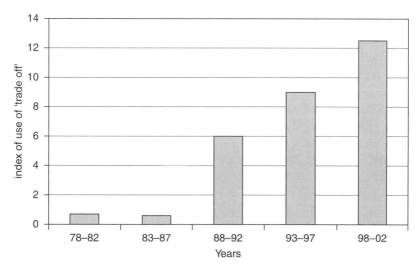

Fig. 4.2: The use of the word 'tradeoff' in the titles of papers in five top ecology journals between 1978 and 2002; data taken from appendix 1 of Nobis and Wohlgemuth (2004). The journals used were *Ecology, Journal of Animal Ecology, Journal of Ecology, Oecologia,* and *Oikos.* All these journals showed an increase in the numbers of papers published per year over this time period, to remove this effect I have divided the number of papers mentioning tradeoff in each time period by the total number of papers published in that period then multiplied by 1,000, to give an index based on whole numbers. Note the increase in explicit use of the term tradeoff since about 1990.

A good ecological example of tradeoffs in action is background matching in cryptic animals (Fig. 4.3). Many species of moth have patterns which match part of their environment: such as the Oak Beauty Moth *Biston strataria* which matches lichen-covered tree trunks (Cott, 1940), while other British moths, such as the Early Thorn *Selenia dentaria* and the Lappet *Gastropacha quercifolia* look like dead leaves (Majerus, 2002). Such insects have a problem, their environment is not just made up either of lichen-covered trunks or nothing but dead leaves, instead it comprises many other potential backgrounds such as green leaves of various shades, lichen-free branches, and colourful flowers. Which background should they match? With a simple mathematical model (Box 4.1) it is possible to investigate the conditions under which an insect should specialize in matching a particular background or have a compromise colouration where it is a less good match to a wider range of backgrounds. There are tradeoffs here, having a good match for one background can limit the range of microhabitats that an animal can safely use. This is important as heterogeneous environments are common in

Fig. 4.3: At its best background matching can be extraordinarily successful. There are two Ptarmigan *Lagopus mutus* in the centre of this photograph of a Scottish mountainside, however, they are so well camouflaged that I considered using another photograph to illustrate this point where the organisms would be easier to see! This species of Grouse increases its background matching by having a white plumage in winter and a grey plumage—as in this photograph—in summer (Cramp and Simmons, 1979), however, against the 'wrong' background it will stand out (e.g. a grey summer bird against a white snow patch or a white winter bird against grey rock and lichen). As such there are potential costs to accurate background matching as well as benefits.

nature, while environments of uniform colour and pattern appear to be much rarer. There is evidence from experiments with captive Great Tits *Parus major* feeding on artificial prey, that under some conditions compromise colouration, which is a reasonable match for more than one background, is a viable answer for cryptic prey living in heterogeneous environments (Merilaita *et al*., 2001), as predicted by the mathematical model outlined in Box 4.1. Using the decathlon analogy, one of the specialist moths will be harder to see than a moth with a more generalized pattern in the specialist's correct microhabitat (it will out-compete the generalist in that particular specialized 'event') but the more generalist moths may out-compete the specialist if the 'game' is played over a wider range of microhabitats (so winning the gold medal for overall performance).

Box 4.1:   A simple model of crypsis in an environment with multiple backgrounds

---

This box is based on the generalized version of the model of Merilaita *et al*. (1999), as described by Ruxton *et al*. (2004). It provides a simple model to illustrate the idea of tradeoffs and when an organism should evolve to specialize and when it should be a generalist.

Consider two backgrounds (a and b), these could be lichen-covered wood and lichen-free wood for the case of a moth that spends time on trees. Let the probability of being viewed by a predator against background 'a' be V_a (similarly V_b is probability of being seen against background 'b'); note that the predator may or may not actually detect the 'moth'. The corresponding probabilities of not being seen against these two backgrounds is C_a or C_b.

Since in this simple example the prey can only be on one of these two backgrounds: $V_a + V_b = 1$ (remember as these are probabilities they must add up to 1)

The overall probability of being detected by the predator (D) is

$$D = V_a(1 - C_a) + V_b(1 - C_b)$$

As any moth must be either seen or not seen by the predator the probability of escaping detection (E) is

$$E = 1 - D \quad (\text{since } D + E = 1)$$

In this example, I am interested in the tradeoff between crypsis against two backgrounds and improved crypsis on one background. In this case C_b is a

Box 4.1: *(Continued)*

declining function of C_a, that is as the chance of the moth not being seen against background 'a' improves, its chance of being seen against background 'b' increases. The maths is slightly more complex than the simple statements of probabilities above, but the important result can be seen in Fig. 4.4 (if you are prepared to take the maths 'on trust' ignore the following equations). Formally:

$$C_b = f(C_a),$$

$df(C_a)/dC_a < 0$ (this is just stating that as C_a increases, C_b decreases)

With a bit more maths it can be shown that E has a maximum or minimum if

$$df(C_a)/dC_a = -V_a/V_b$$

This will be a maximum if

$d^2 f(C_a)/dC_a^2 < 0$ (the curve this produces is shown on the graph)

As can be seen from the graph a series of outcomes are possible, either be a generalist and try and look a bit like both backgrounds or specialize in background 'a' or 'b'. Which outcome is best depends on the actual values for V and C in any given example.

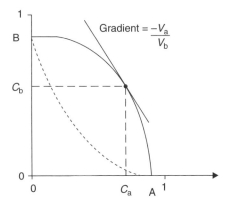

Fig. 4.4: The probability of not being detected when viewed against background 'b' (C_b) as a function of the equivalent probability against background 'a' (C_a). If the function is convex (e.g. the solid line) and has a point where its gradient is $-V_a/V_b$, then a compromise level of crypsis is optimal (it should be a generalist), otherwise the 'moth' should maximize its match to one of the backgrounds (be a specialist). So there will be more cases when it pays to be a specialist so these tradeoffs are likely to lead to increased biodiversity. From Ruxton *et al.* (2004).

This idea of generalist and specialist species is well illustrated by some members of the Corvidae, the bird family containing the crows, jays, and magpies. The Raven *Corvus corax* is widely distributed across the Holarctic (Goodwin, 1986) and seems able to survive in a wide range of climates and habitats. I have watched Ravens on cold wet hillsides in Britain and in the hot desert mountains surrounding Death Valley in California. They do well in wild country where you can walk for hours without seeing another human, but I have also watched them in the middle of alpine villages and on urban fringe refuse dumps in Mediterranean Europe. It is interesting to contrast the Raven with the Nutcracker *Nucifraga caryocatactes*, a dark brown jay-like bird with white spots, this is a specialist largely restricted to a single habitat, montane coniferous forests across Eurasia (Coombs, 1978; Goodwin, 1986). Other specialist members of the Corvidae include the Red-Billed Chough *Pyrrhocorax pyrrhocorax* which uses its slender bill to dig invertebrates—often ants—from the soil, usually in well-grazed areas with a close-cropped vegetation or the closely related Alpine Chough *Pyrrhocorax graculus*, a montane specialist which has been recorded feeding on food dropped by mountaineers near the summit of Everest in the Himalayas (Coombs, 1978; Cramp and Perrins, 1994).

In the context of the thought experiment that I have used to structure this book, an obvious question is what process leads to specialization by organisms (and hence to biodiversity), and would they be expected to operate in all conceivable ecologies? Superficially, it may seem that the best idea is to be a generalist. However, as the decathlon example illustrates, a generalist will usually be outcompeted by a specialist under specific conditions. Generalists win the decathlon because the rules require performance in 10 *different* events. The Raven illustrates this point about competition with specialists, in north America you often see Ravens in desert habitats (MacMahon, 1985; Wilkinson, pers. obs.), however, it is not usually found in the deserts of north Africa or the Middle East where it is replaced by related species from the same genus which appear to be more specialized for desert living (Goodwin, 1986). The athletic analogy is of a great decathlete losing a particular event, say the shot-put, to a lesser decathlete who specializes in that particular event.

The decathlon example illustrates another important point, tradeoffs limit the extent of the generalist strategy. While one person can be reasonably good at all the events in the decathlon, if the sport was broadened to include the skills of a jockey and a sumo wrestler then no one could do well in all of the events. Ravens are found in a wide range of habitats and climates but are absent from others, such as the humid tropics. Tradeoffs make a superorganism or Darwinian Demon, which is good at everything, a figment of the science fiction writers' imagination.

4.4. **Tradeoffs and biodiversity**

The idea that tradeoffs contribute to the generation of biodiversity is not new (e.g. Tilman, 2000), however, it has been given relatively little prominence in academic ecology. For example, I searched for the term 'tradeoff' in the indexes of the current editions of 10 university-level ecology text books (published between 1993 and 2006) and found that only two texts indexed the term (Rickleffs and Miller, 2000; Begon *et al.*, 2006). At least one of the other texts contained a *brief* discussion of the idea without giving it a name (Smith and Smith, 1998). However, naming the idea matters, as Gould (2002, p. 875) wrote 'phenomena without names, and without theories marking them as worthy of notice, will probably not be recognized at all'. The main aim of this chapter is to argue for the central importance of tradeoffs in ecological theory, as such it should feature prominently in the text books. This is an ecological idea 'worthy of notice'!

As argued above, at the most basic level, tradeoffs prevent an organism from being good at everything, and so help rule out persistent planetary ecologies based on a single species. Within the major guilds described in Chapter 3 tradeoffs also make it unlikely that a single species, such as a single taxon of photosynthetic autotroph, will dominate the system. Because of tradeoffs a single photosynthetic taxon cannot perform equally well in a wide range of light levels or, if terrestrial, under very different levels of water availability. As such within any major guild I would expect tradeoffs to lead to a proliferation of species on any planet with life.

Tradeoffs preventing one species out-competing all others, are therefore a good candidate for the most fundamental reason for the development of biodiversity, however, they clearly do not explain all the diversity of life on Earth. For example, dispersal—or lack of it—and other aspects of biogeography play an important role in generating biodiversity on Earth. This is illustrated by the distribution of several families of flightless birds; the rheas (Rheidae) are found in South America, whereas the ostrich (Struthionidae) is now found in Africa but was previously also in parts of Eurasia, and the emus (Dromaiidae) are restricted to Australia (del Hoyo *et al.*, 1992). These distribution patterns (with the proliferation of families) are clearly more to do with accidents of evolutionary history than tradeoffs (see Chapter 10 for a more detailed discussion of historical contingence). However, the convergent characteristics of these three families—along with the Cassowaries (Casuariidae), Kiwis (Apterygidae), and several extinct taxa—is probably partly due to tradeoff constraints as well as shared phylogenies.

Although tradeoffs provide an important theoretical reason for biodiversity, which should operate in all ecologies, the theory is currently qualitative. We have no quantitative theoretical understanding which would allow us to predict the species richness of a planet, even to an order of magnitude! Such predictions may be impossible if historical contingency is important on most planets; however, making these ideas more quantitative would surely be a useful step forward in our ecological understanding.

4.5. **The Gaian effect of biodiversity**

The ecological role of biodiversity, usually defined as species richness for convenience of measurement, became a highly controversial subject at the end of the twentieth century (Grime, 1997). This debate has a longer history (e.g. Elton, 1958); however, the emphasis changed from mid- to late twentieth century. When Robert May (1973) wrote his, now classic, book on *Stability and complexity in model ecosystems* he followed Elton in emphasizing questions in population ecology, such as are populations in complex ecosystems more stable? However, for mathematical convenience May considered small perturbations to equilibrium systems which are probably not representative of many systems in planetary ecology. Recently, the trend has been to start to ask questions about the effect of species richness on ecosystem services (e.g. Naeem *et al.*, 1994; Tilman *et al.*, 1996). As is now well known these questions are not straightforward to answer, for example the more species there are in a system, the more likely it is that one of them will be highly productive, and so give the impression that higher species richness causes higher biomass (an effect described as a 'hidden treatment' by Huston, 1997), so species richness cannot easily be separated from the identity of the species. These questions are now of great importance due to large-scale human-caused extinctions, a problem which has spread from the technical ecological literature into more accessible general texts (e.g. Leakey and Lewin, 1996; Quammen, 1996; Wilson, 1992, 2002). So what are the Gaian effects of biodiversity?

Since plants are basic to most terrestrial food webs, and are easier to experimentally manipulate than most animals, they are a useful group to think about. Consider some measure of the 'performance' of a plant community, such as total biomass. What is the effect of number of species on biomass? One possible answer comes from economics, and is often known as the portfolio effect. The idea is that a cautious investor in a stock market is more likely to sleep soundly at night if they put their money into a portfolio of shares, rather than investing in shares from a single company. Statistically a portfolio investment will show less variance—a lower chance of either very high profits or spectacular losses—than an investment in a single company's shares. This is due to statistical averaging and is effectively an economic application of the idea of tradeoffs, investing in a portfolio trades off the chance of very high returns against a reduced probability of large losses. The idea is illustrated in Fig. 4.5, which shows the effect of investing £7,000 in shares of computing firms quoted on the London Stock Exchange in early 2005. If you had been lucky (or skilful?) and invested all your money in the best performing company, then after a four month period you would have made over £500 profit. However, if you had chosen the worst performing company you would have made a loss of over £1,000! A portfolio investment where your £7,000 was spread equally between seven companies would have given you a much more modest loss. The means and standard deviations of these share prices over 16 weeks are given in Table 4.1, in this case the standard deviations of the worst performing shares and the portfolio are similar (the worst shares were very steady for much of the time then fell sharply), however, the best

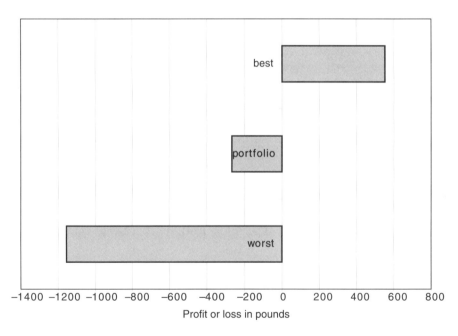

Profit or loss in pounds

Fig. 4.5: The results of investing £7,000 in shares of software and computer companies on the London Stock Exchange for the first 16 weeks of 2005. The portfolio is made up by investing £1,000 in each of seven firms (this made a modest loss). Also shown is the outcome of investing £7,000 in the best performing of these seven companies and in the worst performing. Mean and standard deviations of these share prices (sampled once a week during this period) are given in Table 4.1. Given the poor performance of most of these shares over this period I am pleased that I did this as a thought experiment rather than a real experiment!

performing company shows a much larger variation (if you had sold your shares in this firm at any point over the 16 weeks you would always have made a profit, but the size of that potential profit changed greatly from week to week). The portfolio of seven companies would have given you either a modest profit or a modest loss over the 16-week period depending upon when you sold your shares. This extended example may seem overly detailed to illustrate such a simple idea, however it is likely that this portfolio effect is an important ecological process.

In an ecological context the suggestion is that a species-rich community will also show less variation in measures of performance (such as biomass) for the same reasons (Doak *et al.*, 1998). It is important to realize that this is a statistical relationship, put more colloquially; on average the community will be more stable but there will be exceptions which prove the rule. This is shown by the economic data in Table 4.1 where the lowest standard deviation was actually for the worst performing shares, not the portfolio, for reasons described above. This idea of statistical averaging is most easily understood by considering a monoculture, in such a situation if the single plant species is not growing well then the total biomass of the system is likely to decline—the same would be true of other systems such as a microbial mat (potentially of relevance to a wider range of planets than a 'plant' community). If however, this species is part of a more diverse community and it performs poorly then other species may perform better, and at least in part, take its place; so there may be no decline in biomass. It seems very likely that this statistical averaging (the portfolio effect) will operate in many ecological communities and indeed also at the planetary scale, where total biomass can be very important in the 'physiology' of a planet (see Chapter 6). The extent of this importance may vary between different systems and could depend on the exact nature of the relationship between variance in species abundance and mean species abundance (Tilman *et al.*, 1998). In the simplest case, if the performance of every plant (or microbe or other organism) was positively correlated then the idea would not work, just as a portfolio of shares may not protect you from

Table 4.1: Mean and standard deviations of share prices measured once a week over a 16-week period illustrating the portfolio effect.

	Mean	Standard deviation
Worst shares	35.9	3.48
Best shares	361	9.06
Portfolio	194.9	3.92

See caption of Fig. 4.2 for further details on these data.

financial loss during a stock market crash when almost all shares are declining in value. (For a discussion of other more complex situations when statistical averaging will be a weak effect see the comment by Tilman *et al.* (1998) on Doak *et al.*'s (1998) important original *American Naturalist* paper on the portfolio effect.) An experimental illustration of the portfolio effect was provided by Dang *et al.*, (2005), they studied fungal decomposers in stream microcosms and found that reducing fungal diversity led to increased variability in the rate of decomposition, as would be predicted by a statistical averaging approach.

There are many other reasons why species richness may contribute to the long-term persistence of an ecological system. Once again consider a plant community (or a biofilm if you are a microbiologist), anyone who has spent time describing plant community data—such as the quadrat from the Dolomites described at the start of this chapter—will be familiar with 'lost plants' (*sensu* Janzen, 1986). These are plants which exist in low numbers in a plant community, often as seedlings or stunted individuals of species which are either absent, or very rare, as fully grown individuals in the community. The pine seedling in the Dolomite quadrat is an example of this as the main tree in this part of the forest was Spruce *Picea abies*, the pines dominated the forest higher up the mountainside around the tree line. Janzen called these 'lost plants' by analogy with vagrant animals such as Yellow-Bellied Sapsucker *Sphyrapicus varius* (a north American woodpecker) or the Monarch Butterfly *Danaus plexippus* again from the Americas, both of these species occasionally turn up in Britain after having accidentally crossed the Atlantic.

Such rare organisms are often considered to be of no ecological importance. However, at least in the case of plants, Philip Grime (1998, 2001) has argued that they could be crucial in allowing communities to cope with climate and other changes. Grime called these species 'transients' and suggested that a species rich community with more transients was more likely to contain species with the correct attributes to suit new conditions. As such species-rich communities may be more able to cope with climate change. In this case, species richness increases the probability of a species with particular characteristics being present in the system, this is a real ecological effect and not something which should be written off as an accidental effect—or 'hidden treatment' in an experimental context (cf. Huston, 1997). A very similar idea, predating Grime, was suggested by James Lovelock (1992); in discussing the role of biodiversity in Daisyworld models (described in Chapter 6) he wrote, 'destroying biodiversity will reduce the reservoir of apparently redundant, or rare species. Amongst these may be those able to flourish and sustain the ecosystem when the next perturbation occurs'. The tendency to dismiss species as redundant, as far as ecosystem function is concerned, may be partly a product of thinking about them at the wrong time scale. On the 3-or 4-year timescale of many PhD studies these species may be doing nothing of importance, but on a longer time scale with climate and other changes, this may not be the case.

A problem with these ideas is that they make predictions about the behaviour of communities over longer time scales (perhaps tens to thousands of years), this makes them hard to study. An obvious approach is to reconstruct past changes using palaeoecological methods. The relevance of this approach to 'lost plants' is illustrated by data from a study of a peatland at Astley Moss in

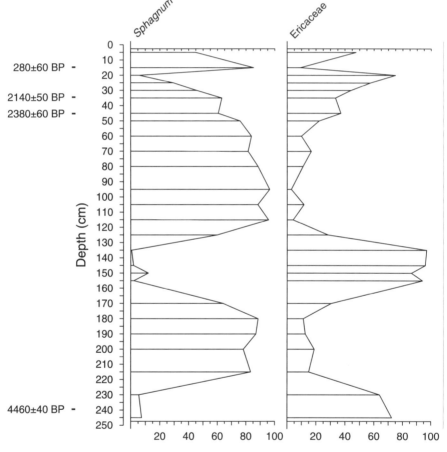

Fig. 4.6: Macrofossil diagram for Astley Moss (part of the Chat Moss peatland complex in northwest England described in more detail in Section 8.1) showing the abundance of *Sphagnum* moss and ericaceae (the heather family) remains in the peat. This record covers approximately 5,000 years, the radiocarbon dates given down the left-hand side of the diagram are in 'radiocarbon' years. Calendar years tend to be a little older so that the date from near the base of the core is calibrated to between 5,292 and 4,881 calendar years ago. Note how when *Sphagnum* is common ericaceae are rare and vice versa. A more detailed macrofossil diagram for this site—showing a wider range of taxa—is given by Davis and Wilkinson (2004).

northwest England by Steve Davis and myself (Davis and Wilkinson, 2004). We reconstructed the bog vegetation over the last 5,000 years by looking at the remains of plant leaves and roots preserved in the peat. Over time the vegetation alternated between communities dominated by Bog Moss *Sphagnum Spp.* and the ones dominated by heathers (*Calluna vulgaris* and other species). When *Sphagnum* dominated the community, heathers formed a minor component of the vegetation and vice versa (Fig. 4.6), so as conditions changed species that had been rare in the vegetation took over as dominants maintaining vegetation cover at the site. Similar patterns could also be seen in the abundance of testate amoebae whose shells are preserved in peats along with the plant remains, over time different species were dominant, with pronounced changes in the community approximately 2,500 years ago, probably due to climate change. Clearly the analysis of palaeoecological data has promise in investigating ideas about the long-term role of rare species in communities; indeed, this literature is rich in examples similar to our Astley Moss study (an especially good example being the highly detailed pollen diagrams of Turner *et al.*, 1993).

In an illuminating discussion of ecosystems from the perspective of the engineers' concept of redundancy, Shahid Naeem (1998) pointed out that a basic principle of reliability engineering is that 'the probability of reliable systems performance is closely tied to the redundancy in its design'. So ecosystem function should be more 'reliable' with more species. This engineering perspective is consistent with the ideas of Lovelock and Grime described above; however, it illustrates another great problem with such theories, namely producing a sensible definition of reliability in this context. More modelling may at least lead to sharper definitions in this area.

4.6. **Overview**

Tradeoffs are a fundamental aspect of biodiversity as they prevent one or two species from monopolizing the planet. As such they allow speciation, so that on any hypothetical planet with life we would expect to see a range of taxa. Well-known ecological concepts such as the niche only make sense in the context of the more fundamental idea of tradeoff, if organisms could be good at everything (supergeneralists or Dawinian Demons) then the niche concept would be useless. As such, tradeoffs are a more fundamental idea than the niche or the classical mechanisms of speciation, as tradeoffs are required for these other processes to operate. The resulting biodiversity will have a positive Gaian effect, that is it will tend to make an ecological community, or a planetary ecosystem, more stable than if it was composed of a smaller number of species. Biodiversity does not evolve to help stabilize the system (except in the limited sense that taxon poor systems may be more prone to extinction), it is an inevitable by-product of tradeoffs and other processes such as geographical isolation. One potentially important way to think about the Gaian effect of biodiversity is the idea of the 'portfolio effect' from economics, although other ideas, such as Grime's 'transient species' are also important in this context. In the Universe, stability is likely be an important by-product of biodiversity which should emerge on any planet with life.

5. *Ecological hypercycles—covering a planet with life*

5.1. **Darwin's Earth Worms**

Charles Darwin's final book addressed *The formation of vegetable mould through the action of worms with observations on their habits* (Darwin, 1881). In it he describes the feeding habits of earth worms, writing, 'They swallow an enormous quantity of earth, out of which they extract any digestible matter . . . They also consume a large number of half decayed leaves of all kinds' (Darwin, 1881, p. 35). He also describes the effects of earth worms on the soil, for example, he discusses the amount of soil overlying several British archaeological sites which he largely attributes to the action of worms over the generations (Fig. 5.1). He also describes how this soil formation by worms improves growing conditions for plants, 'Worms prepare the ground in an excellent manner for the growth of fibrous-rooted plants and seedlings of all kinds' (Darwin 1881, p. 309).

The science historian Eileen Crist has argued that Darwin's worm book, although not an anticipation of Gaia, can be 'read, without strain, as a geophysiological thesis' (Crist, 2004, p. 165). She points to the emphasis on how the worms modify their environment in a way that is beneficial to future generations of worms and an implicit description of the positive and negative feedback loops between the worms and their biotic and abiotic environments. For these same reasons Darwin's worm book also came close to describing the idea of an ecological hypercycle.

5.2. **Hypercycles in ecology**

Darwin's description of earth worm ecology could be summarized by the following relationships:

$$\text{Plants} \rightarrow \text{Worms}$$
$$\text{Worms} \rightarrow \text{Plants}$$

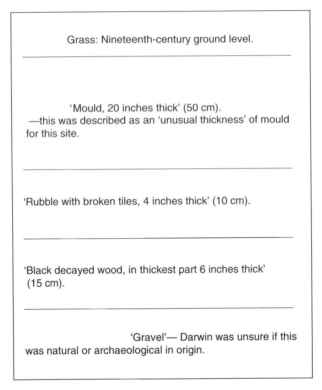

Grass: Nineteenth-century ground level.

'Mould, 20 inches thick' (50 cm).
—this was described as an 'unusual thickness' of mould
for this site.

'Rubble with broken tiles, 4 inches thick' (10 cm).

'Black decayed wood, in thickest part 6 inches thick'
(15 cm).

'Gravel'— Darwin was unsure if this
was natural or archaeological in origin.

Fig. 5.1:   Diagrammatic representation of a soil profile from nineteenth century excavations at the Roman site of Silchester in southern England. This was a large Roman town which was extensively excavated during the 1800s, partly because of the accessibility of the remains, as there had been relatively little post Roman building on the site (de la Bedoyere, 2002). These data were supplied to Charles Darwin by the Rev. J.G. Joyce, one of the excavators. By 'Mould' Darwin meant good quality soil which he assumed had been processed by passing through the guts of worms. This diagram is based on fig. 11 in Darwin (1881), with measurements in centimetres and other annotations added. Text in quotation marks is taken directly from Darwin's diagram or his accompanying text. Darwin used this and several similar archaeological studies to illustrate the rate of soil formation, over a time scale of less than 2,000 years, which he attributed to earth worm activity. These studies showed biology having a significant effect on its physical environment.

Where the arrows show a positive effect of one group on the other. Although the following summary would appear obvious to a twenty-first century reader it is never explicitly described by Darwin.

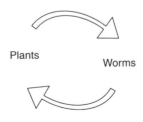

Plants

Worms

This diagram shows a positive feedback loop and is an example of a simple hypercycle (Eigen and Schuster, 1977; Maynard Smith and Szathmáry, 1995), where the rate of replication of each group (plants and worms) is an increasing function of the size of the preceding group. So more plants means more worms (more food), and more worms means more plants ('better' soil for growth). One of the main applications of hypercycles in biology has been in considering the auto-catalytic effects in cycles of self-replicating molecules during the origin of life, however, they are clearly also relevant to ecosystems (Maynard Smith and Szathmáry, 1995). For example, the dung-based system described in Section 3.2 can be represented as such a cycle.

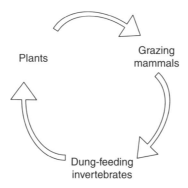

If plants increase then there is more food for the mammals (be they deer, antelopes, or elephants) which feed on them. If the mammals increase in numbers then there is more food for the guild of dung-feeding invertebrates and an increase in the numbers of these invertebrates speeds up the decomposition of the dung, so making more nutrients available for the plants.

One of the difficulties with such ecological hypercycles is that they are usually a great simplification of the real situation, in reality additional arrows often need to be added. Consider clearings in temperate forest; a biome which covered much of Britain before forest clearance for agriculture and other purposes (Bennett, 1989; Rackham, 2003). Mammals such as various deer species and the now extinct wild cattle *Bos primigenius* would have fed on the lush plant growth in the clearing, so tending to halt succession back to woodland by making it hard for trees to recolonize (Buckland and Edwards, 1984; see also Innes and Blackford, 2003, for indirect evidence of a high biomass of grazing mammals in such clearings). So if the plant guild in the hypercycle is restricted to plants of open conditions (i.e. excluding the trees), then there should be an additional arrow going from grazing mammals to plants, with plants benefiting grazers and grazers directly benefiting the plants through halting succession—as well as indirectly benefiting them through the dung route. Complications such as these make it harder to formally model these ideas in the context of complex ecosystems. However, the concept of hyper-cycle is still a useful aid to ecological thought, even if difficult to describe in a quantitative manner.

Although a simplification, the idea that life can have an auto-catalytic qual-ity is of great importance in understanding the ecological history of a hypo-thetical planet (or real planet such as the Earth). Once autotrophs and decomposers have evolved, as discussed in Chapter 3, then the simple hyper-cycle *autotroph* ↔ *decomposers* will exist. Note that this is really just a generalized version of the *plant* ↔ *earth worm* hypercycle I described above. This has implications for the spread of life on any planet on which it evolves—or on which it arrives if you favour panspermia rather than an *in situ* origin of life.

5.3. **Covering a planet with life**

These auto-catalytic hypercycles, such as the *autotroph* ↔ *decomposer* exam-ple described above could quickly lead to life covering much of a planet, as the presence of life catalyses the production of more life. Indeed Nisbet and Fowler (2003, p. 22) suggest that 'on a geological timescale, once the first cell had replicated, all habitats on the planet would be immediately filled'. Why should this be the case? As well as hypercycles another mechanism is the well-known potential for geometric growth in unconstrained populations, when the net reproductive rate (R_0) is greater than one, which also leads to the same conclusion; namely life quickly covering the planet. Equations of the form

$$N_{t+1} = R_0 \, N_t$$

are common in ecology textbooks (where N_t = population at time t and N_{t+1} = population at the later time of $t+1$), so once life starts reproducing there is the potential for it to quickly reach huge population sizes. This large amount of life can then be subdivided into different species by the processes described in the previous chapter. Similar arguments (along with those discussed in the next chapter) are also the reason why Lovelock (2000a) argued that planets will tend to have either lots of life or no life at all.

This conclusion, of life covering a planet in a geological instant, clearly requires the habitat for life to be widespread. On Earth it is usually assumed that early life could survive in aquatic habitats and since the planet had widespread oceans (although some of these could have been frozen) it could quickly spread (Nisbet and Sleep, 2001; Nisbet and Fowler, 2003).

5.4. Why would persistent restricted ecologies be unlikely?

An alternative possibility is that life may survive on a planet in only a few locations. There are at least two main reasons why this is unlikely, one is because of environmental stochasticity tending to cause extinction (described below), although the other is that it would be unable to play an important role in the 'physiology' of its planet (the subject of the next chapter) and as such be unable to contribute to planetary regulation. The first of these reasons will be familiar to most ecologists as it is effectively part of the 'small population paradigm' of conservation biology applied at a planetary scale. One of the reasons why conservationists worry about small populations is that they are prone to extinction brought about by demographic and environmental stochasticity (Lande, 1993; Caughley, 1994; Simberloff, 1998a).

Demographic stochasticity is when chance fluctuations in demographic parameters (e.g. birth and death rates, sex ratio, population growth rate) in a small population can lead to extinction. Consider sex ratio in the simple situation where each generation of an organism does not overlap—for example, the adults produce eggs or seeds and then die off before the next generation is ready to breed (many insects and annual plants show this behaviour). It is intuitively obvious that if there was only one member of the next generation then it must either be male or female, giving the trivial result that the probability of all members of that generation being the same sex would be 1 (i.e. 100%) if the generation size was one. If this species has an equal probability of producing offspring of either sex and there were two individuals in the next generation then the probability of them both being the same sex would be 0.5 (logically the probability of the second individual being the same sex as the first must be 0.5). It is possible to calculate probabilities for more complex cases using the following relationship (Simberloff, 1998a):

$$\text{Probability of all individuals being the same sex} = 2^{1-N}$$

where N = number of individuals in a given generation. The results of applying this relationship to several small population sizes are shown in Table 5.1. As can be seen at very small population sizes (e.g. 5) there is a real possibility of all individuals being of the same sex by chance. By the time the population size is 10 individuals this is quite unlikely (but would still be expected to occur once or twice in every 1,000 cases), while by the time the population consists of 100 individuals it is exceedingly unlikely. Although this particular example only applies to sexually reproducing organisms, these results are typical for demographic stochasticity, in that these processes apply mainly to very rare organisms. As such these ideas are much less relevant to microbial ecology than they are to large multicellular organisms such as the Californian Condor *Gymnogyps californianus*. For example, while there were only about 20 California Condors in the wild in the early 1980s before they were taken into captivity (del Hoya *et al.*, 1994), there can be around 4×10^4 cells/ml of the marine prokaryote *Prochlorococcus spp* (Whitmann *et al.*, 1998) and even the much larger testate amoebae species can reach densities of 2,300 individuals/cm^3 in peatland soils (Tolonen *et al.*, 1992). So the very low population sizes of relevance to demographic stochasticity are not usually relevant to microbes and as such not relevant to the early history of life on a planet; apart from a possibly exceedingly brief period at the very start of life when for a geological instant, population sizes may have been almost non-existent.

Ideas of environmental stochasticity and catastrophes are very different, being potentially much more relevant to arguments that persistent restricted ecologies do not survive for geological periods of time. Good illustrations of the importance of these processes in conventional conservation biology are given by several bird species. An example are the frigatebirds (Fregatidae), there are currently five species, three of which have quite widespread distributions in the tropics, whereas two are restricted to breeding on a single island: the Ascension Frigatebird *Fregata aquila* and the Christmas Frigatebird *F. andrewsi* (del Hoyo *et al.*, 1992). The population of Ascension Frigatebirds (Fig. 5.2) is estimated to be around 9,000–12,000 birds (Bourne and Simmons, 1998). Although this is

Table 5.1: Probability of all individuals in a generation being of the same sex for a species which produces males and females with equal probability and has non-overlapping generations.

Number of individuals	Probability all of one sex
2	0.5
5	0.062
10	1.9×10^{-3}
100	1.6×10^{-30}

Fig. 5.2: Ascension Frigatebirds *Fregata aquila*, this photograph shows approximately 0.1% of the total population of this species. Such relatively rare species can be prone to extinction through environmental stochasticity.

quite a small population it is still much too large to be easily put at risk by demographic stochasticity. However, it only breeds on Ascension Island and in recent decades its breeding has largely been restricted to the small Boatswain Bird Islet, off the east of the main island, because of feral cat predation (these cats have now been exterminated). As such the whole population of birds could be wiped out by a single catastrophe, such as unusual weather events or outbreaks of disease.

Another well-known example of an island bird which did suffer extinction after humans arrived on its island home, was the Dodo *Ralpus cucullatus*. This was only found on Mauritius; so once it was extinct on this one island it was globally extinct, therefore its very limited range was a key factor in its extinction. A particularly good example of environmental stochasticity is the demise of another flightless bird, the Great Auk *Alca impennis*. This penguin-like sea bird used to be relatively common in the north Atlantic, but became easy prey to human hunters as their remote breeding colonies became more accessible. One of its last main breeding colonies was the island of Geirfuglasker off the coast of Iceland, and the catastrophic destruction of this island by volcanic activity in 1830 hastened the demise of what was by then a rare bird (Fuller, 2000); earlier in its history when it was more widely distributed this disaster would have been much less of a threat to the survival of the whole species.

As these avian examples demonstrate, organisms that have very restricted distribution are vulnerable to catastrophes—both physical such as volcanic eruptions or biological such as disease outbreaks (although disease may seldom be enough to cause extinction by itself; see Section 3.4). Early in the history of the Earth there was a period of potentially catastrophic asteroid and cometry impacts, often referred to as the 'Heavy Bombardment', lasting from approximately 4.4 to 3.8 billion years ago (Grieve, 1998; Wells *et al.*, 2003). This may have prevented the survival of any life on Earth at this time, certainly any life which was widely spread over the planet was much more likely to survive; this is also true for more recent meteorite impacts, glaciations and other catastrophes. This is an important reason why life is only likely to show long-term survival if it covers much of a planet, although it should be pointed out that in Lande's (1993) models, populations with high growth rates were surprisingly resistant to environmental stochasticity, so even geographically restricted microbial populations may survive all but the most dramatic planetary catastrophes.

5.5. **The end of life on a planet**

If the idea that spatially restricted ecologies are unlikely to survive is true, then it raises interesting questions about the end of life on a planet. A good way to think about this problem is to use Mars as an example and make the big assumption that this planet had widespread microbial life in the past. Today, Mars gives a lifeless impression, its atmosphere—very different from that of the Earth— shows that biology must at least be exceedingly rare. Before spacecraft had ever landed on the planet James Lovelock and Lynn Margulis (Lovelock and Margulis, 1974; Margulis and Lovelock, 1974) had used arguments about the nature of the Martian atmosphere to predict the lack of life. However, if life had been widespread at some point in the past then there would have been large numbers of microbial cells—for example there are currently an estimated 4–6×10^{30} prokaryote cells on Earth (Whitman *et al.*, 1998). It may be very difficult to bring about the extinction of such a colossal number of microorganisms by processes other than the catastrophic destruction of the whole planet, indeed analogies from Earth-based microbiology suggests that it will probably take a long time for all microbial cells to die. On Earth large numbers of microbes live in the surface of rocks in seemingly inhospitable deserts, this includes both hot dry sites such as in the Californian desert (Kuhlman *et al.*, 2005) and cold dry sites such as in the Antarctic dry valleys (Wynn-Williams, 1996). Both of these are habitats which have attracted the interest of researchers curious about the possibility of life on

Mars. Earth-based microbiology also suggests that microorganisms (especially prokaryotes) are able to survive in a dormant state for a very long time, some papers have suggested survival for up to 600 million years, although these studies are very controversial and rather lower figures are probably more realistic (Box 5.1). In theory microbial communities should survive better in a dormant state on a dying planet as they will be less likely to be attacked by other organisms than on an ecologically highly active planet such as the Earth. An ability to remain dormant, potentially for hundreds of millions of years, along with an ability for small numbers of cells to survive worsening conditions in habitats such as

---

Box 5.1: **Survival of dormant microorganisms**

How long can bacteria and other microbes survive in a dormant state? There is a long history of studies which claim to have isolated viable microbes—usually bacteria—from geological sediments. All such studies have to confront two very difficult problems, first, avoiding contamination with modern microbes and secondarily determining that the microbes were incorporated into these sediments at the time of formation, rather than at a later date. In addition, the accurate dating of the sediments can also prove difficult. This is not an easy area of research, although the results are potentially fascinating!

Kennedy *et al.* (1994) tabulated the published evidence for long-lived microorganisms. When their data for claimed survival of 1 million years or more is plotted against publication date an interesting pattern emerges (Fig. 5.3). With the exception of the pioneering work of Lipman (1928), little was published until the middle of the twentieth century, when a series of claims for microbial survival of hundreds of millions of years were published; several suggesting survival from over 600 million years ago. Such extreme claims vanished from the literature in the later twentieth century, more recently papers typically argued for survival of from between 1 and 200 million years. There are two main explanations for this lack of older ages since the 1960s. One possibility is that improved experimental methods may have greatly reduced the chance of contamination, while the other explanation is that the more robust refereeing of papers has made it far harder to publish such claims (even if they are correct!).

Vreeland *et al.* (2000) published evidence for 250-million-year-old bacteria from an inclusion in a salt crystal, however even this detailed study was challenged, with doubts over the age of the salt inclusion (Hazen and Roedder, 2001). Although these controversial claims of survival for hundreds of million of years are the most relevant to arguments for the long tail to life on a dying planet, records of survival for hundreds or thousands of

Box 5.1: (*Continued*)

years are more common and usually less controversial (Postgate, 1990; Kennedy *et al*., 1994; Miskin *et al*., 1998); however, these are still subject to the same problems of contamination and dating.

An alternative approach is to try and directly calculate the survival half life for a population of dormant bacteria. For example, some bacteria can survive radiation doses of up to 2,000–3,000 grays (Madigan *et al*., 2000), however under most conditions on Earth this could be achieved from background radiation in a few million years at most. Similarly, a dormant microbe would be subject to a slow death from the simple diffusion of its cell contents over time. Considerations such as these make life spans of hundreds of millions of years *in a dormant state* appear unlikely. However, these are probabilistic processes (very like the idea of radioactive half life), since microbial populations can be unimaginably large, a lucky few cells could survive for much longer than the vast bulk of the population.

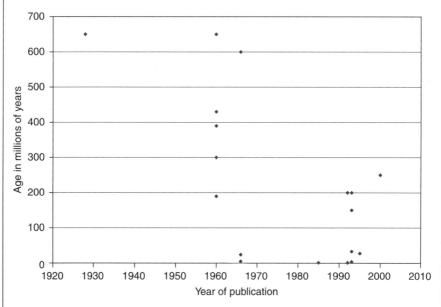

Fig. 5.3:   Claimed age of dormant microbes plotted against year of publication of study, for studies claiming ages of 1 million years or more published during the twentieth century. Data taken from table 1 of Kennedy *et al*. (1994), with the addition of the studies of Cano and Borucki (1995) and Vreeland *et al*. (2000).

within rocks, would lead to life having a long 'tail' on a dying planet; petering out over hundreds (or thousands?) of millions of years. However there is a problem with such arguments, if the cells are restricted to surviving in a dormant condition they will be unable to repair the unavoidable damage they will accumulate over time and so eventually die out, although it is very difficult to attach numbers to such a process (Box 5.1). Certainly the potential 'long tail' to microbial life on a planet would make it impossible to identify the moment of death of all life on a planet. As Oliver Morton (2002, p. 297) wrote of Mars, 'if you search for individual living organisms, it is almost impossible to answer in the negative; to say categorically that there is no living thing on Mars would require knowing the planet with an intimacy that is hard to imagine'. However, long before the last cell died a planet could be declared ecologically dead, its biogeochemical cycles having become effectively just geochemical cycles, of no professional interest to the ecologist. The exceptions to a long time span for the extinction of life on a planet would be catastrophic events, such as the massive asteroid or cometry impacts suffered by the Earth in its early history or a potentially lethal flux of cosmic rays from a nearby supernova—in this case nearby may mean around 30 light years or less (Narlikar, 1999). Even in these cases it may be very difficult to kill every last microbe living deep within a planet's rocks.

5.6. **Overview**

Ecological hypercycles are auto-catalytic processes by which different organisms (or guilds) improve each others environments. A good example being autotrophs producing material of use to decomposers and the decomposers releasing nutrients which are reused by the autotrophs. Such an auto-catalytic process clearly has a positive Gaian effect and is one of several reasons for expecting life to quickly cover most of a planet's surface once it has evolved. Environmental stochasticity makes it unlikely that ecologies restricted to a small area of a planet will survive for a geological period of time. However, once widespread, life may have a long drawn-out end on a dying planet, with some cells surviving for at least millions of years after life last played any role in the planet's physiology (i.e. after biogeochemical cycles returned to being mere geochemical cycles).

6. *Merging of organismal and ecological physiology*

6.1. **From beavers to planetary ecology**

The beaver is a classic example of an organism which modifies its environment for its own benefit. There are two species of beaver in the genus *Castor*: the North American beaver *C. canadensis* and the Eurasian beaver *C. fiber*. Both species modify their environments through the building of dams and other engineering works, with similar behaviours and densities of dams recorded from both North America and Western Europe (Coles, 2001); although some studies suggest that the European species may be less prone to large engineering works (reviewed by Rosell *et al.*, 2005). These activities of beavers affect streams in at least six distinct ways (Coles, 2001): (1) altering channel dimensions, (2) favouring the development of multiple channels, (3) changing water velocities, (4) increasing the volume of water in stream channels, (5) changing the pattern of sedimentation, and (6) increasing organic input to streams (cf. Hodkinson, 1975). Clearly the activity of beavers does not just affect the plants they eat but also has important effects on their physical environment.

The extent to which beavers construct these engineering works depends upon the local hydrology and other factors. They usually construct dams when the water is shallow, and by raising the water level they produce ponds which allow them both to swim to their food sources and store food supplies under water. In addition, they can create canals which allow them to float branches and so more easily move them around (Macdonald *et al.*, 1995; Rosell *et al.*, 2005). Therefore by modifying their environment—'ecological engineering' in the terminology of Jones *et al.* (1994, 1997)—they provide themselves with a better environment. However, their engineering also modifies the environment for other species; in general, species which like wetter conditions can do well as a by-product of beaver activity, especially if they prefer still rather than flowing water (e.g. I have watched large numbers of dragonflies breeding in beaver ponds). Those that like dryer conditions or flowing water will often suffer as a result of the rodent's ecological engineering (e.g. you often see dead trees standing in beaver ponds which have been killed by the raised water levels). One example of the potential effect of beavers on other species is the suggestion that beavers have a positive effect on the environment for otters *Lutra spp* in both Europe and North America. Beaver ponds provide a good hunting environment for otters and in winter they also benefit from the beavers' habit of maintaining ice free areas as breathing holes and for access to the water, this has led to the suggestion that recent increases in otter numbers in parts of North America may be owing to increased populations of beavers (Rosell *et al.*, 2005).

The previous chapter argued that life will very quickly colonize a whole planet, and an inevitable consequence of this is a large biomass of life on any inhabited world. As such life will not only influence its local environment but also has the potential to cause important effects at the planetary scale. These influences could be through direct adaptations to modify the environment for its own benefit (e.g. a beaver dam), or accidentally through by-products (e.g. CO_2 from respiration or O_2 from photosynthesis). As the intention of this book is to primarily focus on concepts, I will first discuss a simple but informative model illustrating the potential for life to influence planetary ecologies before discussing some of the mechanisms by which this can happen. I will then argue that such an approach forces us to give biomass—rather than species richness—much more attention than has traditionally been the case in ecology and that the mass ratio theory should be at the centre of planetary, as well as plant community, ecology.

6.2. **Daisyworld**

Open a textbook on mathematical ecology (e.g. Case, 2000) and along with the logistic growth curve, you will find that the Lotka–Volterra models of interacting populations have formed a starting point for much of the twentieth-century theoretical ecology. These models are typical of much of mathematical ecology in modelling the interaction between organisms without considering feedback between the organism and their abiotic environment. Although the majority of modern ecologists probably only know of Alfred Lotka through his eponymous equation, it was not typical of his approach to ecological theory. Lotka—working in the 1920s—regarded the organic and the abiotic aspects of the world as a single system in which it was impossible to understand the working of any part of the system without considering the whole (Kingsland, 1995). This led him to suggest that modelling the whole system (biotic and abiotic) would prove simpler than attempting to understand an unrealistically isolated fragment; an approach sympathetic to that taken in this book. However, mainstream theoretical ecology has usually proceeded by considering organisms as isolated systems. Although this may be a reasonable simplification for modelling *some* ecological systems of limited spatial and temporal scale it is certainly less appropriate for planetary scale systems.

Since the early 1980s an alternative approach to ecological modelling has developed that is much closer to Lotka's philosophy and illustrates the potential global importance of the presence of life on a planet. However, most of these so-called Daisyworld models have been published in non-ecological journals and as such have had little impact on mainstream ecological thinking. For example out of 90 citations to one of the founding papers of this approach (Watson and Lovelock, 1983) listed by *Web of Science* (in January 2006), only 9 were from ecological journals. The origin of Daisyworld models lies in an attempt by James Lovelock to counter claims of teleology levelled against his Gaia hypothesis (see Lovelock 2000b for a personal account of the history of the model). In classic versions of Daisyworld, a cloudless planet is the home of two kinds of plants— conventionally called daisies (Watson and Lovelock, 1983; Lenton and Lovelock, 2001). One type is dark coloured (ground covered by it reflects less light than does bare ground) and the other is light coloured, reflecting more light than bare ground. These daisies are usually referred to as 'black' and 'white', although the key point is the way in which their albedo differs from unvegetated ground. The growth rates of these daisies are assumed to be a parabolic function of temperature. Black daisies absorb more light energy than do white ones and so become warmer. The temperature of the planet Daisyworld is governed by the input of stellar energy and the albedo of the planet, so life and the physical environment are therefore coupled in the model (see Box 6.1 for a summary of the mathematics underlying the model).

To understand the behaviour of a simple Daisyworld model, consider a planet that is slowly receiving more stellar energy as its star warms; this is true for the Earth as the sun was fainter in the past (Sagan and Chyba, 1997). The planet reaches a temperature that enables daisies to grow and their seeds to germinate; however, as conditions are cool the black daisies grow better than the white ones because their dark colour allows them to absorb more stellar radiation and so heat up. The black daisies will therefore come to dominate the initial community. As these dark plants spread, they warm the planet, providing a positive feedback which enhances the spread of life on the planet. At high temperatures, the white daisies have a selective advantage over the black daisies, and so their population expands which tends to cool the planet by increasing its albedo. Eventually only white daisies are left and are finally killed off as the temperature rises even further.

Box 6.1:   **An introduction to the mathematics of Daisyworld**

The original mathematical description of Daisyworld was given by Watson and Lovelock (1983), for a good recent summary see Lenton and Lovelock (2001). Most investigations of Daisyworld have relied on simulations; however, analytical solutions to Daisyworld models are described by Saunders (1994) and Sugimoto (2002).

The population dynamics of the plants are described using the epidemic model used by Carter and Prince (1981) to model aspects of plant biogeography. So that the equation for black daisies is

$$d\alpha_w/dt = \alpha_w(x\beta_w - \gamma) \tag{1}$$

Each daisy population is described as its fractional cover of the planet α (as Daisyworld is a mathematical planet it is assigned an area of 1 for convenience). β is the growth rate per unit time and area, whereas γ is the death rate per unit time. The subscripts 'w' and 'b' signify white and black daisies. Studies by Maddock (1991) suggested that Daisyworld was not particularly sensitive to the type of population model used. 'x' is the fraction of the planet's surface not covered by daisies, so that

$$x = 1 - \alpha_w - \alpha_b \tag{2}$$

In standard Daisyworld models the plants have an optimum growth temperature of 22.5°C and limits of tolerance to temperature of 5°C and 40°. The growth rate (β) of each daisy type is a parabolic function of their local temperature (T_i):

$$\beta = \max(0, 1 - ((22.5 - T_i)/17.5)^2). \tag{3}$$

The effective temperature of the planet T_e is given by the following energy balance equation:

$$\sigma(T_e + 273)^4 = SL(1 - A_p). \tag{4}$$

where σ is the Stefan–Boltzmann constant, S is a constant energy flux (so that $S/\sigma = 1.68 \times 10^{10}$ K^4), and L is a dimensionless measure of the luminosity of Daisyworlds star (273 appears in equation (4) to keep all temperatures in degree centigrade, this being the freezing point of water on the kelvin scale). A_p is the albedo of the planet, which is given by

Box 6.1: *(Continued)*

$$A_p = \alpha_g A_g + \alpha_b A_b + \alpha_w A_w \tag{5}$$

where A_g is the albedo of bare ground, A_b albedo of black daisies, and A_w white daisies—on Daisyworld $A_b < A_g < A_w$. The local temperature of black daisies can be determined from the following equality which relates local temperatures to planetary temperatures:

$$(T_b + 273)^4 = q(A_p - A_b) + (T_e + 273)^4. \tag{6}$$

where $q = 2.06 \times 10^9$ K^4 and is a measure of insulation between the various regions on the planets surface. A similar equation can be written for white daisies.

This original version of Daisyworld is zero dimensional, that is there is no explicit representation of space. More spatially complex versions of Daisyworld now exist; examples include a one-dimensional model with latitudinal variations in stellar radiation (Adams *et al.*, 2003) and a two-dimensional cellular automata model with a curved planetary surface (Ackland *et al.*, 2003).

The result of the daisies' behaviour is a considerable degree of planetary temperature regulation. Over a substantial range of luminosities the daisies act as a negative feedback on planetary temperature, this is illustrated in Fig. 6.1 which shows results from a slightly more complex model with three different colours of daisies. The reason for this negative feedback, and the resulting temperature regulation, is that any increase in a daisy of a particular colour tends to force the planetary temperature towards conditions which are less good for the growth of that type of daisy. For example an increase in black daisies will increase planetary temperature, eventually forcing the temperature into a range where black daisies grow less well than the cooler white daisies; this leads to a reduction in black daisies and a decrease in temperature. An important aspect of conventional Daisyworld is that the plant's adaptation (colour) has the same effect on the plant (warming or cooling) as it does on the planet, and obviously this will not always be the case in the real world (but see Seto and Akagi, 2005). However, Daisyworld illustrates the potential for biology to play an important role in regulating conditions on a planet, but for this to happen it requires life to cover much of the planet's surface. If the daisies were restricted to a small fraction of Daisyworld then their effects on planetary albedo would be negligible and their

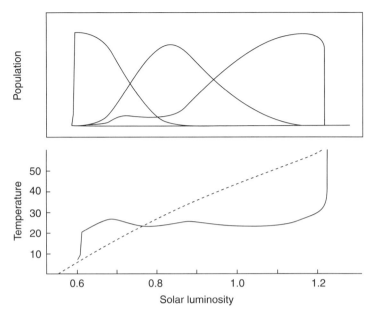

Fig. 6.1: Climate evolution on a three species Daisyworld, from Lovelock (2000a). In this version a grey daisy species was included as well as the classic black and white daisies. These grey daisies were introduced to attempt to counter the criticism that 'cheats' which did not have the cost of making pigments could undermine the system. This idea that cheats can destroy otherwise cooperative interactions between different species is a common theoretical problem in evolutionary ecology (Yu, 2001). Lovelock (2000a) added a minor cost to pigment production for black and white daisies. The upper graph shows the population sizes of the three daisies (black peaking at lower solar luminosities, then grey, then white). The lower graph shows the temperature on Daisyworld with plants (solid line) and without any life (dashed line). Clearly the grey daisies did not undermine the temperature regulation, although one possible criticism is that all three Daisies have seedbanks of infinite life span, so however well the cheat did it could not cause the extinction of the other species.

ability to regulate its temperature would be lost. Here is another potential reason for expecting life on any planet with a persistent ecology to be a planet-wide phenomenon as it allows life to be involved in regulating processes of planetary significance.

Daisyworlds have attracted a lot of attention and show surprisingly interesting behaviour for such simple models, thus far, however, they have mainly been investigated by people from a physics or climatology background and there is scope for more ecological investigations of Daisyworld (limited examples so far include Lovelock, 1992; Harding and Lovelock, 1996; Harding, 1999; Cohen and Rich, 2000: Seto and Akagi, 2005). There is also an interesting attempt by Pujol (2002) to develop a maximum entropy production (MEP; see Section 2.4) approach to Daisyworld, he showed that this increased the ability of the biota to regulate temperature compared to more conventional Daisyworld models. However, in this book I am primarily using Daisyworld as a simple conceptual model to illustrate the potential for life to be involved in the regulation of planetary processes if it covers a large enough area of a planet.

6.3. **Examples of the role of life in planetary physiology on Earth**

Is there anything operating like the black and white daisies on Earth? One possibility is the Boreal forest, this covers some 15.8 million km^2 with another 3.3 million km^2 of coniferous forest in temperate mountain areas (Archibold, 1995). Much of the area covered by this great expanse of trees has lying snow, especially during winter. However, these coniferous tree species often shed their snow so that they remain dark coloured, compared to the high albedo of the surrounding snow covered ground (Fig. 6.2). Even where snow survives on the tree canopy, multiple reflections within the canopy scatter the light rather than reflecting it, so the vegetation still has a lower albedo than a snow-covered ground surface (Harding and Pomeroy, 1996). This raises the possibility that the boreal

Fig. 6.2:   On a small scale this photograph of the English Peak District in winter illustrates Richard Betts (2000) point about the Boreal Forest. The snow shedding shape of these coniferous trees making them black 'daisies' in a white winter landscape.

forest may actually warm high latitude areas, so potentially helping to maintain a climate suitable for trees. An interesting question which follows from these observations is, what is the net climatic effect of boreal forest in the context of global warming? We normally think of forests as having a beneficial effect by acting as carbon sinks and so potentially reducing global warming, but the net warming effect of creating a lower albedo at higher latitudes would tend to counteract the 'cooling' effect of carbon sequestration. Modelling suggests that 'in many boreal forest areas, the positive forcing induced by decreased albedo can offset the negative forcing that is expected from carbon sequestration' (Betts, 2000).

Clearly the Boreal forest could potentially act like a black daisy, what about white ones? Possible examples of a white daisy are some species of marine algae. In this interesting example the 'white daisies' are not the algae themselves but an indirect by-product, namely clouds. Dimethylsulphoniopropionate (DMSP) is found at high concentrations (100–400 mmol^{-1}) in the cells of many marine algae, including coccolithophorids and dinoflagellates. DMSP has a range of possible functions in algal cells including osmoprotection and as an antioxidant (Simó, 2001; Sunda *et al.*, 2002). DMSP breaks down to dimethylsulphide (DMS), which is the most common form of volatile sulphur in the oceans and a major source of cloud condensation nuclei. This gives the possibility of a Daisyworld-like mechanism, often referred to as the CLAW hypothesis after the initials of the authors of the original paper suggesting this mechanism in *Nature* (Charlson, Lovelock, Andreae and Warren, 1987); this paper is now a citation classic with over 1,360 citations in *Web of Science* (as of January 2006). Warmer conditions potentially favour algal growth so leading to more DMS and so more cloud and therefore cooler conditions with less DMS production; this is a negative feedback system. Just as when Daisyworld gets warmer the white daisies expand and so cool the planet, on Earth as algae expand they make more cloud and cool the planet. Over the last two decades, the explosion in research in this area, much of it triggered by the CLAW paper, has complicated this relatively simple idea. For example, it is now possible that under glacial conditions the mechanism may operate as a positive feedback loop, making the planet even colder. This could happen because many marine algae are limited by nutrient levels in the ocean (especially Iron). In the arid conditions of glacials much more dust may blow into the oceans adding nutrients and so increasing algal growth and DMS production therefore increasing cloud cover (reviewed by Welsh, 2000). An additional complication is that colder conditions can cause thermal stratification of the oceans, isolating the surface layers from more nutrient-rich waters below. More work is needed to sort out the complexities of this system—for example Simó (2001) lists a selection of important questions concerning the ecology and evolution of this system which need addressing.

Irrespective of the overall sign of this feedback loop the DMS system illustrates microbial life having a large, and potentially regulating, effect on a planet's climate. In addition this DMS plays a major part in the sulphur cycle by transferring large quantities of this element from the oceans to the land, indeed it was this postulated role in cycling sulphur which originally caused James Lovelock to look for DMS production by marine algae in the early 1970s (Lovelock, 2000b; Simó, 2001). Through the production of DMSP the physiology of algal cells plays an important role in the physiology of the planet, by cycling sulphur and affecting the climate.

The bog mosses *Sphagnum spp* also have much in common with the plants of the planet Daisyworld. These mosses are characteristic plants of many peat bogs and thrive in wet conditions. *Sphagnum* can hold up to 20 times its own weight in water with the aid of special empty 'hyaline' cells which allow it to act like a sponge. As such the moss makes its own local environment wetter, and so more suitable for it, while also making the whole bog ecosystem wetter (Porley and Hodgetts, 2005). Just as the model 'daisies' have the same local and global effects (warming or cooling) *Sphagnum* has a similar local and ecosystem-wide effect with water content (and also with pH, increasing the acidity both locally around its cells and more widely across the bog system). *Sphagnum* is a supreme ecological engineer, as Porley and Hodgetts (2005, p. 361) have written '*Sphagnum* does not just live in the bog: it *is* the bog', helping to control the moisture content and acidity of the whole system, while often physically constructing the bog from its own peat forming remains.

Life has been implicated in the regulation of many other aspects of the Earth system; examples include CO_2 and O_2 levels in the atmosphere (for O_2 see Chapter 7 and for CO_2 Chapter 8). Another important example is the nutrient balance of the Earth's oceans. Classic studies by Alfred Redfield in the 1930s identified a clear empirical relationship between the physiology and biochemistry of plankton and the physiology and chemistry of the oceans (Redfield, 1958; Falkowski and Davis, 2004). Redfield pointed out that the nitrogen : phosphorus ratio of many plankton was approximately 16 : 1 and that the N : P ratio of seawater was also very similar. This strongly suggested a major role for plankton in regulating ocean chemistry through a range of processes including the acquisition of nutrients by plankton, the formation of new planktonic biomass and the remineralization of this biomass by bacteria (Falkowski and Davis, 2004). Our lack of a detailed understanding of these systems is highlighted by the recent discovery that the rate of nitrogen fixation in the open ocean appears to be much higher than previously thought (Capone *et al.*, 2005), prior to this work it had been thought that most nitrogen in the ocean's euphotic zone came from the mixing of nitrate from deeper waters rather than from planktonic N-fixation. Indeed it seems that nitrogen fixation by photosynthetic bacteria plays an important role in the negative feedback loops which help maintain the ocean at Redfield ratios. If the rate of phosphorous input (via rivers from weathering on land) rises then the N-fixation allows plankton numbers to expand to use this increased phosphate thus bringing the N : P ratio back to around the Redfield values (Tyrrell, 1999; Lenton and Watson, 2000a). The Redfield ratios are a particularly spectacular example of organismal physiology playing a major role in the physiology of the whole planet, but to have this effect requires a high biomass of plankton. Indeed, these ratios appear to be average values which emerge from the different physiologies or a range of phytoplankton growth strategies rather than a global optimum value for all plankton (Arrigo, 2005), this makes their persistence even more remarkable.

6.4. The importance of biomass, an illustration from the Earth's past

Just over 65 million years ago at the end of the Cretaceous a dramatic event apparently triggered one of the five great mass extinction events known from the fossil record. These Phanerozoic mass extinctions (there were no doubt others during the Proterozoic and Archaean) are of more than historical interest as we may currently be causing a 'sixth extinction' through our own actions (Leakey and Lewin, 1995). There is now a high level of agreement that a large meteorite impact was associated with the end Cretaceous extinction although volcanism, especially in the Deccan Traps in India, may also have played a role (for general reviews of this mass extinction see Glen (1994) for a historian's perspective; Alvarez (1997) for a popular account, and Pope *et al.* (1998) for references to the more recent technical literature).

The probable impact site for the meteorite has been identified in the Yucatan Peninsula in Mexico, in this area the carbonate rocks were anhydrite ($CaSO_4$) rich, and an impact here would have caused large amounts of sulphur dioxide (SO_2) to enter the atmosphere on impact. This increase in atmospheric SO_2 would have been detrimental to many life forms as it would have caused extensive acid rain; however, it would also have acted as an ultra-violet light shield, helping to compensate for damage done to the ozone layer by the impact (Pope *et al.*, 1998; Cockell and Blaustein, 2000). A combination of paleobotanical, isotope and modelling work has suggested that the terrestrial biomass recovered extremely quickly (possibly in less than 100 years), while the marine system took some 3 million years to return to its pre impact productivity (Beerling *et al.*, 2001a; Lomax *et al.*, 2001). Thus biomass recovered on a much faster scale—even in the oceans—than was required for the evolution of replacement species, therefore *productivity recovered much faster than biodiversity*. This conclusion is highly relevant to the discussions of the role of biodiversity in academic ecology, which has tended to focus on much shorter time scales.

How did the terrestrial biomass recover so quickly? Studies of fossil leaves suggest that the impact caused high levels of atmospheric CO_2, considered to be due to an instantaneous transfer of carbon from sediments to the atmosphere caused by the impact (Beerling *et al.*, 2002a). Modelling (Lomax *et al.*, 2001) suggests that these raised CO_2 levels would have allowed high photosynthetic productivity. This is particularly interesting as paleobotanical data suggests that the initial recovery of terrestrial biomass was quick and involved relatively few species, initially dominated by ferns (Beerling *et al.*, 2001). Although these fossils suggest that the recovery was rapid, it is the models which give it a time scale of decades (such precision is usually impossible to get from the record in the rocks). However it is possible that unusual conditions, such as very high CO_2 levels, may not be required for rapid recovery. For example, high biomass forests have returned to much of eastern North America since the decline of farming in the nineteenth- and early twentieth century (Foster, 1992) and an entirely man-made forest has formed on the previously treeless Ascension Island over a similar time scale (Wilkinson, 2004a). Given a suitable climate and soils, high biomass vegetation can easily develop in a time scale of 100 years or less if ecologically appropriate plant species are available. It is relevant that the environmental effects of the supposed impact may not have been as severe as some previous authors have suggested (e.g. Wolbach *et al.*, 1988; Alvarez, 1997); for example, studies of fossil charcoal suggest that there was not a global destruction of vegetation by fire (Belcher *et al.*, 2005) so that the meteorite is only likely to have sterilized the seed bank local to the impact site. If Belcher *et al.* (2005) are correct then previous evidence of 'soot' from the end of the Cretaceous (e.g. Wolbach *et al.*, 1988) may have come from hydrocarbons in rocks at the impact site rather than the burning of global biomass. This lack of major global fire storms would clearly make the analogy with the modern forests described above seem more realistic.

These palaeobotanical data are of interest in the context of Grime's (1998) suggestions on the functioning of modern plant communities. He argued that 'the extent to which a plant species affects ecosystem function is likely to be closely predictable from its contribution to the total biomass'; an idea he calls the 'mass ratio' hypothesis. In other words, it is the total biomass (normally made up of a very limited subset of species) which governs most ecosystem services, the majority of rare species contribute very little. The end Cretaceous extinction event is consistent with this in that a vegetation which had lost many of its species apparently returned to a large biomass, with its associated role in carbon cycling, and so on, possibly on a time scale of decades. As such it was biomass not species richness which was crucial to the recovery, although it is possible that the more species rich the vegetation was before impact, the greater the chance of it containing species suitable for post-impact conditions, and this may be the most important role for biodiversity when viewed at a planetary scale (see Section 4.5).

6.5. **Biomass and Gaia**

The arguments I have developed above make a plausible case that life on a planet will be widespread and therefore have a substantial biomass (Chapter 3) and that because of this biomass the products of these organisms' physiology will play a large, and potentially regulating role, in the physiology of the planet (this chapter). It is clear that there is the potential for systems involving life to have a self-regulating character, but what about abiotic systems? The natural nuclear reactors which existed 2 billion years ago at what is now Oklo in Gabon, West Africa are a good illustration of a self-regulating natural system probably not having any significant biological involvement and hence showing that life is not necessary for regulation. These geological nuclear reactors comprised uranium ore in sandstone rocks overlying impermeable granite and were probably naturally moderated by water, if the reactors ran too fast this water turned to steam so reducing its ability to slow the neutrons and prevent them from being absorbed by naturally occurring ^{238}U (Maynard Smith and Szathmáry, 1995; Meshik, 2005). This controlled the rate of the nuclear reactions in a manner analogous to that used in nuclear power plants. In this case there is a possibility that bacteria, some of which can sequester uranium, may have been involved in the initial formation of these ore deposits (James Lovelock, pers. comm.); however, even if this was so the likelihood is that the regulation of these reactors was largely abiotic. Since regulation does not require biology, a key question is, are there any theoretical reasons to expect systems involving life to be more likely to be self-regulating than solely abiotic systems?

In the context of the Earth a plausible answer to this question is that Earth systems do appear to have regulated conditions within a life-friendly range over geological time; however, this is accidental, with no reason to expect the same to happen on any other planet with life. Several people raised this possibility in the context of Gaia theory at the end of the 1990s (Lenton, 1998; Watson, 1999; Wilkinson, 1999a); however, it is Andy Watson who has taken these arguments most seriously (Watson, 1999, 2004). The basic idea is easily explained and has much in common with the anthropic principle in astronomy, where the presence of astronomers clearly implies aspects of the nature of the Universe (Carr and Rees, 1979; Hoyle and Wickramasinghe, 1999b). As I have previously written, 'Any planet which is home to organisms as complex as James Lovelock and Lynn Margulis must have had a long period of time during which conditions were always suitable for life, and thus must give the impression of regulation for life-friendly conditions even if the persistence of life was purely a matter of chance' (Wilkinson, 2004b, p. 72).

As long as these arguments are based on a single example—the Earth—it is impossible to rule out an explanation based entirely on chance, however, such an explanation does seem rather unlikely. If life evolves very rarely—or just once—on a planetary surface then it is remarkably unlikely that chance would provide a self-regulating system. If one assembles biospheres in such a random way then it would seem likely that there would be many more ways of producing non-regulating systems than regulating ones. If we assume that life arises on planets reasonably frequently then there is the possibility of a selection process, so that only planets which develop self-regulating systems—even if this is 'by chance'—are likely to survive for long as a living planet. However, it seems likely that the full explanation involves special properties of planetary systems involving life.

The most important, and provocative, part of Gaia theory is that a planet's environment is regulated 'at a habitable state for whatever is the current biota' (Lovelock, 2003, p. 769). Imagine a planet where the climate and chemistry are currently in a suitable state for its prevailing life forms—an easy task as we live on such a planet. If the activities of life were tending to force this system towards an uninhabitable state, for example through the production of some metabolic by-product, then as the planet approached this uninhabitable state these organisms would increasingly struggle and so the forcing would be reduced (an alternative is that natural selection may produce something which consumes the by-product; for example, see Section 3.1 on lignin). These in-built regulating mechanisms, which by definition work to keep the system in life-friendly conditions, are not present with purely abiotic forcings. With the evolution of a significant biomass of more complex organisms the range over which this regulation happens will tend to decrease; for example, the upper temperature limit for prokaryotes is higher than for single-celled eukaryotes which are less temperature sensitive than vascular plants (Table 1.1). This is the significance of the phrase 'whatever the current biota', in Lovelock's definition of Gaia cited above. Just as Grime (1998) argues that the high biomass of dominant species largely controls the behaviour of a plant community—the mass ratio hypothesis—so in a similar manner the dominant life forms (by biomass, rather than by number of species or individuals) may control the properties of a planet's biosphere—at least until the rise of an intelligent species with an industrial economy. These Gaian mechanisms can be thought of as loading the dice in favour of the long-term survival of life once it has started on a planet. In the 'just chance' anthropic view the survival of life is like rolling a 6 with a fair dice (although the probability of 'success' is likely to be much less than 0.17), while Gaian mechanisms load the dice to make a 6 much more likely—although probably not inevitable. Therefore, key questions for a Gaian view of planetary ecology are what are the mechanisms? and how strongly do they load the dice?

6.6. **Overview**

All organisms modify their environment to some extent, either as an evolved adaptation (e.g. beaver dams) or through by-products of their activity (e.g. oxygen release by photosynthesis). Daisyworld models illustrate the theoretical possibility that these modifications could lead to planetary regulation for life-friendly conditions, but also illustrate the need for substantial biomass for this to occur. Examples of possible regulatory processes on Earth include the production of DMSP by many marine phytoplankton and the stability of the N : P ratio of sea water. Crucially for life to have these effects it must have a considerable biomass, so that its products can significantly influence a planetary system. In academic ecology, the recent concentration on biodiversity may have tended to obscure the fundamental importance of biomass—which will often be dominated by a limited number of species. This 'mass ratio' effect is important both in the ecological community and on a planetary scale. Once biomass is sufficient for the merging of organismal and ecological physiology then the planet will tend to show some level of regulation for life-friendly conditions, as such this merging—and its associated high biomass—has a positive Gaian effect. The magnitude of this expected effect on any planet is currently an open question, it could range from a slight increase in the probability of life surviving to a much tighter regulation reminiscent of homeostasis in an organism. The size of this effect is currently a key unknown in both Earth-based environmental science and astrobiology.

7. *Photosynthesis*

7.1. Quantification and mysticism in the seventeenth century

Imagine trying to understand how a plant grows if you had not been taught any biochemistry. Animals clearly eat to gain resources, but plants? An obvious guess is that they are eating soil, but a famous seventeenth-century experiment shows this to be wrong. The Flemish chemist and medic Joan Baptista van Helmont (1579–1644) planted a willow sapling in a pot and found that after five years of watering the tree had increased in weight by 164 lb (74.4 kg: and that was excluding 4 years worth of fallen leaves); however, the soil had lost negligible weight. His involvement in this long-term experiment was perhaps encouraged by the fact that he had been forbidden to travel by The Inquisition (Ball, 2005b). Van Helmont's interpretation was that the plant's increase in weight must have come from the water, the correct answer that it had largely come from the 'air' would have seemed incredible. The construction of a tree from elemental water made sense to Van Helmont who was a vitalist and saw life in everything, even the rocks—a conclusion not dissimilar to a Gaian perspective which stresses the heavily intertwined nature of the biotic and abiotic environment, but arrived at for entirely different reasons. Indeed he regarded water as a primary element which could form plants and much else. While a proponent of the experimental method he was also very at home with a magical view of the Universe—part 'scientist' and part mystic (Porter, 1997; Henry, 2002). Although this seems odd from a twenty-first century perspective it was not an unusual world view for the natural philosophers of the seventeenth century; an obvious example being Isaac Newton's interest in alchemy (Dobbs, 1982).

The ability of photosynthetic organisms to use such common resources as CO_2, water, and light to make carbohydrates has big implications for Earth-based ecology and raises interesting questions about the likelihood of photosynthesis evolving on any planet with Earth-type life. The release of oxygen as a waste product of oxygenic photosynthesis has been a key factor in the Earth's ecological history and the evolution of its atmosphere.

7.2. **The diversity of photosynthesis on Earth**

When most ecologists think of photosynthesis they tend to picture a green plant, such as Van Helmont's willow sapling; however, on a global scale this is rather misleading. On Earth, about half of the primary production is produced by marine microbes (Falkowski, 2002; Arrigo, 2005), our knowledge of these has increased tremendously over the last few decades but it has not yet attained much prominence in the education of most ecologists.

One of the key figures in the ecology of marine plankton during the first part of the twentieth century was Alister Hardy. In his classic mid-century account of the natural history of plankton (Hardy, 1956) he gave a snapshot of the state of knowledge at the time. In describing photosynthetic phytoplankton he mainly wrote about the larger taxa such as diatoms, but briefly mentions that there are 'vast numbers' of smaller eukaryotes which pass 'through the mesh of the finest net we can use' (Hardy, 1956, p. 49). Of prokaryotes he wrote, 'smaller still are the bacteria . . . at present very little is known about their occurrence in the plankton' (Hardy, 1956, p. 50). Since Hardy's book it has become apparent that very small phytoplankton are enormously important in both marine and fresh-water habitats—the marine taxa being the most important in the Earth System because of the vastly greater extent of the oceans. These microbes are often referred to as the picophytoplankton and can be defined as photosynthetic O_2-evolvers which can pass through 2 μm diameter pores (Raven, 1998), much smaller than the 'finest bolting cloth' used by millers for sieving flour which Hardy (1956) recommended for making plankton nets.

Some of the most important members of the picophytoplankton are the cyanobacteria, one of the key taxa being *Prochlorococcus* spp. Although not described until 1988 this genus dominates primary production in the tropical and subtropical oceans and must be a good candidate for the title of the commonest organism on Earth (Fuhrman, 2003; Sullivan *et al.*, 2003). It has a diameter of about 0.6 μm and the smallest genome of any photosynthetic organism yet studied (1,716 genes in the high light adapted ecotype: Rocap *et al.*, 2003). Along with the slightly larger (0.9 *μ*m) *Synechococcus* spp these plankton are major contributors to primary production. In the context of my discussion of the potential role of parasites in population regulation and evolu-tion (Sections 3.4 and 3.5) it is interesting to note that these picophytoplankton are the hosts for a range of viruses (Culley *et al.*, 2003), indeed viruses appear to be an important and understudied aspect of marine ecology (Culley *et al.*, 2003; Suttle, 2005).

On land we often describe the main biomes based on a mix of climate and the dominant types of photosynthetic plants (often trees), such as tropical forests, tropical savannas, Mediterranean ecosystems, and temperate forests (e.g.

Archibold, 1995). However, these biomes also contain photosynthetic microbes and in some systems with few true plants they can be very important. For example, Hodkinson *et al.* (2004) studied the invertebrates along a chronosequence created by a retreating glacier on the Arctic Island of Svalbard. When I examined sediments from their sites under the microscope it was obvious that there were large numbers of cyanobacteria, even in sediments which had only been exposed for a couple of years. Indeed, these microbes dominate many terrestrial and freshwater polar habitats and play an important role in the carbon and nitrogen cycles at these sites (Vincent, 2000).

So far I have attempted to give an overview of the taxonomic and structural diversity of photosynthetic organisms, from bacteria to large trees; however, there is also important biochemical diversity, for example the range of types of carotenoids is much greater when we consider all photosynthetic organisms rather than just green plants (Box 7.1). There is also diversity in the basic chemistry of photosynthesis, the summary equation for oxygenic photosynthesis is

---

Box 7.1: **Carotenoids**

Carotenoids are yellow, orange, or red pigments found in all photosynthetic organisms, with a much greater diversity being exhibited by the photosynthetic prokaryotes than plants. They perform several useful functions in true plants, mainly in the folding and maintenance of various photosynthetic proteins and in photoprotective mechanisms which allow excess light energy absorbed by chlorophylls to be dissipated as heat, and to deactivate potentially damaging reactive oxygen 'species' (Förster and Pogson, 2004). In addition, they can absorb light and so potentially act as accessory light-harvesting pigments, but this function is probably not very important in true plants since they absorb light in the same wavelengths as the chlorophylls. However, in some photosynthetic bacteria carotenoids absorb light over wavelengths different from the bacteriochlorophylls (Madigan *et al.*, 2000). This raises interesting questions about their evolution: while most plant scientists are used to thinking of them as primarily photoprotective, could they have first evolved wholly or partly as photosynthetic pigments?

Carotenoids are most obvious to people living in the temperate zones during autumn (fall) when they are one of the main sources of autumn leaf colour, indeed the xanthophylls (one of the main classes of carotenoids) were named after their contribution to autumn colour—xanthos = yellow and phyll = leaf (Förster and Pogson, 2004).

well known and will have been familiar to most readers of this book since some time in school science classes.

$$CO_2 + 2H_2O \rightarrow Light \rightarrow (CH_2O) + O_2 + H_2O$$

Most green plants achieve this by so called C_3 photosynthesis, where CO_2 is fixed into the three carbon molecule 3-phosphoglyceric acid (commonly shortened to PGA). The most common variant on this is C_4 photosynthesis where the initial products are fixed as four carbon organic acids which are moved to where they are needed for photosynthesis within the cell, this reduces problems with photorespiration (Mooney and Ehleringer, 1997). The other common variant on plant photosynthetic pathways is crassulacean acid metabolism, here the separate carboxylation takes place in the same location within the leaf but separated in time (Fig. 7.1).

This biochemical variation is, however, within just one type of photosynthesis, namely oxygenic photosynthesis. However some microbes (such as green sulphur

Fig. 7.1: CAM plants comprise approximately 7% of terrestrial plant species and are distributed across some 33 families (Cushman, 2004). The Joshua Tree *Yucca brevifolia* (photographed in the Mojave Desert, California, USA), is one of the largest plant species to use the CAM variant of oxygenic photosynthesis, this allows it to open its stomata at night to fix CO_2 which is then stored for use in photosynthesis during the day—when light is available. Night time stomatal opening allows the plant to reduce water loss associated with acquiring CO_2, indeed it can provide CAM plants with a five to ten fold increase in water use efficiency relative to C_4 or C_3 plants (Crawford, 1989; Cushman, 2004).

bacteria) carry out anoxygenic photosynthesis, where the source of reductant is the hydrogen in H_2S rather than the H_2O familiar from oxygenic photosynthesis (Fenchel and Finlay, 1995). This can be summarized as

$$CO_2 + 2H_2S \rightarrow Light \rightarrow (CH_2O) + 2S + H_2O$$

Note that oxygen is not a by-product of this process, this has considerable implications for planetary ecologies.

7.3. **Photosynthesis and the Earth system**

It is now clear that the chloroplasts in plant cells were originally derived from prokaryotes (Margulis, 1971; Margulis and Dolen, 2002), therefore questions about the early history of photosynthesis on Earth are really questions about the origin of photosynthesis in bacteria. There has been a general assumption by many workers that anoxygenic photosynthesis predates oxygenic photosynthesis (Fenchel and Finley, 1995), conclusions which have been supported by more recent molecular analysis (Xiong *et al.*, 2000).

The evolution of anoxygenic photosynthesis on a planet is ecologically important as it provides a source of energy for organisms which does not just rely on pre-existing organic chemicals. Circumstantial evidence for anoxygenic photosynthesis has been described from 3,416 million year old rock in South Africa (Tice and Lowe, 2004). However, the rise of oxygenic photosynthesis is probably an even more important major transition in the history of the Earth's environment as it led to a greatly increased supply of free energy for life (Lenton *et al.*, 2004). Aerobic food chains can potentially support many more trophic levels than anaerobic ones (Fig. 7.2). Since size is a major factor in structuring many food chains, longer food chains make possible the abundance of macroscopic animals which feature so prominently in many ecology textbooks. In a detailed study of published food webs Cohen *et al.* (1993) found that about 90% of the feeding links between animal species with known sizes had a larger animal consuming a smaller prey. Obvious exceptions to this general rule are most parasites, also many whales and several species of shark are surprisingly larger than one might expect given the fact that they eat plankton! But in general, multicellular predation may require an oxygen-rich atmosphere for its evolution, for both the physiological and ecological reasons outlined above. These arguments suggest that oxygenic photosynthesis may be intimately connected with a planet's biomass and hence the likelihood of the merging of organismal and ecological physiology (Chapter 6). Anoxygenic photosynthesis will increase the system's energy source thereby potentially increasing the biomass; however,

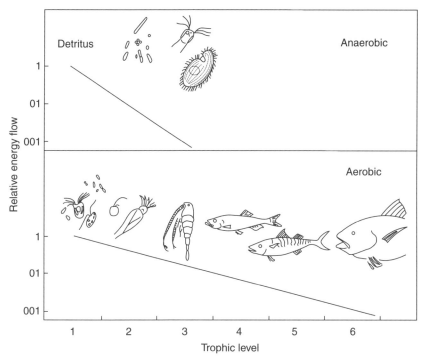

Fig. 7.2: Oxygen should allow longer food chains on a planet. In extant anaerobic fermentation on Earth the energy yield is about 25% that of aerobes. Assuming a growth efficiency of 10% for anaerobic food chains and 40% for aerobic ones, then a similar amount of energy will support a much longer food chain in an aerobic system (from Fenchel and Finlay, 1995).

oxygenic photosynthesis—if it leads to an oxygen-rich environment—sets the stage for a far larger planetary biomass.

7.4. **Oxygen and the Earth system**

There are two main possibilities for the source of oxygen in the Earth's atmosphere: the abiological photo-dissociation of water molecules or oxygenic photosynthesis (Catling and Claire, 2005). The mechanism for the abiological photo-dissociation of water is simple, ultra-violet light is energetic enough to break apart water molecules into hydrogen and oxygen, the hydrogen is light enough to escape the Earth's atmosphere into space, leaving the spare oxygen behind. However, by the middle of the twentieth century it was apparent that this process could not explain the observed levels of oxygen in Earth's atmosphere. The problem is that ozone (O_3) is formed as the levels of diatomic oxygen (O_2) build up in the atmosphere and this ozone shields water molecules from the incoming ultra-violet rays, so reducing oxygen formation (Berkner and Marshall, 1972). This ozone problem means that photosynthesis must be the main source of oxygen in the atmosphere. As James Lovelock pointed out in the 1960s, oxygen in the Earth's atmosphere is a tell-tale indicator of life on our planet, especially as it is coexisting with methane which is not a chemically stable situation (Lovelock, 1965; Lovelock and Margulis, 1974). As far as we are aware photosynthesis is the only process which causes the photo-dissociation of water using visible light—as opposed to ultra violet light (Sagan *et al.*, 1993).

Consider this pairing of the summary equations for oxygenic photosynthesis and aerobic respiration.

$$CO_2 + 2H_2O \rightarrow (CH_2O) + O_2 + H_2O$$
$$(CH_2O) + O_2 \rightarrow CO_2 + H_2O$$

Biochemically, aerobic respiration is the reverse of photosynthesis. Think about a dead plant (or cyanobacterium), to a large extent its structure has been made from inorganic carbon fixed by photosynthesis. As the dead plant is consumed by detritivores (or the live plant by herbivores) the organic carbon is converted back into inorganic carbon (CO_2) and re-enters the atmosphere. Therefore, when considering the amount of atmospheric oxygen, the key factor is the amount of biological material buried in sediments, as this is not respired and therefore there is some oxygen left over from its formation by photosynthesis (see Holmes (1944) for an early textbook account; Berner (1999, 2004); Lenton (2001) for more recent discussion). Indeed this is why marine phytoplankton are so important in contributing oxygen to the atmosphere as a small fraction of their biomass is buried in marine sediments, whereas on land most biomass is respired, so using up the oxygen it produced during its lifetime (Kasting and Siefert, 2002). The claims often made by less scientifically informed environmentalists that the rain forests are the 'lungs of the world' because of their claimed oxygen production is incorrect because most of the luxuriant plant growth is respired on its death unless washed down rivers into marine sediments or added to peat deposits.

The date of the first oxygenic photosynthesis and the timing of the rise in oxygen in the Earth's atmosphere are the subject of considerable dispute (Nisbet and Fowler, 2003). Although the rise in oxygen cannot have happened before the evolution of photosynthesis it is likely that it may not have happened until well after the first photosynthetic oxygen production. If the first free oxygen reacted with chemicals on the early Earth (such as reduced ferrous iron) this would have delayed the rise in atmospheric oxygen. As described in Section 1.4 many structures classically interpreted as very early microfossils have the appearance of cyanobacteria (reviewed by Schopf (1999, 2000)), but there are now considerable doubts about their biological origin. If these 'fossils' are indeed cyanobacteria this raises the possibility of an oxygen rich atmosphere 3.5 billion years ago. One problem with this comes from the faint young sun paradox, which points out that the sun was much fainter in the past. To prevent the freezing of all the water on the early Earth some kind of greenhouse effect is usually suggested. Although Sagan and Chyba (1997) suggested an ammonia greenhouse, most workers currently assume that CH_4 and CO_2 would have been the main greenhouse gases (Nisbet and Fowler, 2003), much of this methane could have been of biological origin (Box 7.2), clearly a methane greenhouse would be incompatible with high levels of oxygen because of the following reaction:

$$CH_4 + 2O_2 \rightarrow CO_2 + 2H_2O$$

Many of these problems would be at least partially solved if it turns out that oxygenic photosynthesis evolved rather later than many people had assumed during the 1990s, indeed recently dates of 2.7 billion years are now more commonly being suggested for the first cyanobacteria (Des Marais, 2005). However, the rise in the Earth's oxygen levels appears to have been a slow process, for example biomarkers ('fossils' of molecules such as carotenoids—see Box 7.1) from 1.6 billion years ago are consistent with a marine microbial ecology with purple sulphur bacteria (Chromatiaceae) and green sulphur bacteria (Chlorobiaceae), this hints that oxygen was well below modern levels at this time (Brocks *et al.*, 2005). Currently, the best geological evidence suggests a first increase in atmospheric oxygen on Earth around 2.4–2.3 billion years ago with an additional important increase some 0.8–0.6 billion years ago (Catling and Claire, 2005).

While the history of oxygen levels in the Archean and Proterozoic are problematic, the levels in the geologically recent past (the last 350 million years) have also been contentious. It seems clear that over this period oxygen levels have shown a certain amount of stability (Berner 1999; Lenton, 2001; Catling and Claire, 2005), however there are disagreements over its extent. The classic palaeoecological approach to try and set limits to the oxygen levels in the Earth's atmosphere was developed by Watson *et al.* (1978), they pointed out that since the development of the first forests there was a continual history of both extensive terrestrial vegetation and charcoal in the fossil record. The conclusion they drew from this was that oxygen levels had never been low enough to prevent forest fires—or there would be gaps in the charcoal record—and never so high that wet vegetation could burn uncontrollably—otherwise forests could not have existed (Fig 7.3). They attempted to quantify these oxygen levels by burning paper strips (proxies for vegetation) in different oxygen levels, concluding that oxygen abundance in the atmosphere must have stayed between 15% (dry paper would not ignite) and 25% (wet paper would burn). While the lower limit has been generally accepted, the upper limit of 25% has been controversial (e.g. Robinson, 1991;

Box 7.2: **Methane**

Currently on Earth there are two main types of microbial methanogenesis (Fenchel *et al.*, 1998). The first are microbes which break down acetate according to

$$CH_3COOH \rightarrow CO_2 + CH_4$$

Examples include *Methanosarcina* and *Methanosaeta*. The second are methanogens (e.g. *Methanobacterium* and *Methanococcus*) which use hydrogen as an electron donor, so that

$$4H_2 + CO_2 \rightarrow CH_4 + 2H_2O$$

The evolution of methanogens is usually assumed to predate photo-synthesis, a conclusion with some molecular support (Gribaldo and Philippe, 2002). One possibility is that life on the early Earth starts in rare hot habitats (e.g. hydrothermal vents). Over time microbial methane production would create a strong greenhouse effect since CH_4 is an important greenhouse gas— approximately 20 times more potent than CO_2 (Fenchel *et al.*, 1998). This would 'improve' the climate from the perspective of Earth-type life, allowing it to spread more widely on the planet (Nisbet and Fowler, 2003), because of the fact that the faint young sun paradox suggests that conditions on the early Earth may have been cold. If this scenario turns out to be correct, then CH_4 played an crucial role in making the early Earth suitable for a planet wide ecology. This view of Archean CH_4 is consistent with both Gaia theory (Lovelock, 1979, 2003) and the ideas of Niche Construction of Odling-Smee *et al.* (2003).

The main natural source of CH_4 in the current atmosphere is usually con-sidered to come from microbes in anoxic environments. However, just as I was about to submit this book to the publishers the situation was complicated by a 'Letter' to *Nature* describing the production of CH_4 by terrestrial plants under oxic conditions (Keppler *et al.*, 2006). If confirmed, their measure-ments suggest that up to 30% of the CH_4 in the Earth's atmosphere could be derived from living and dead plant matter (without the aid of anaerobic microbes). It appears that this methane production had been missed by past workers because of the difficulty in measuring these small CH_4 fluxes from plants in an atmosphere containing CH_4 from other sources—Keppler *et al.* approached this by making measurements in CH_4 free air. As methane is a potent greenhouse gas these new results could turn out to be very important in the context of global warming. As I write this the mechanism for this CH_4 production is unknown.

Fig. 7.3: Fire and vegetation: this photograph shows clouds of smoke billowing across Yosemite Valley in California, USA, from a large forest fire. During Earth's history the higher the atmospheric oxygen content, the greater the chance of large fires destroying significant amounts of vegetation. On the modern Earth, fire consumes so much plant material that it can be usefully compared with herbivorous animals, indeed Bond and Keeley (2005) describe fires as effectively a 'global herbivore' in the title of a paper which argued that the significance of fire in global ecology justifies much more ecological attention than has previously been the case. It may be relevant that these authors have institutional addresses in South Africa and California, both areas rich in Mediterranean systems which are often very fire prone (Archibold, 1995), a scientist working in undamaged rainforest would probably draw rather different conclusions about the importance of fire under current atmospheric levels of O_2. Within the Yosemite National Park fire is now viewed as a natural part of the ecology and smaller fires are often left to run their course.

Berner, 1999). New combustion experiments by Wildman *et al.* (2004) using more realistic plant material (wooden dowels and needles from White Pine *Pinus stobus*) suggested that oxygen levels of 35% would not lead to uncontrollable wildfires destroying all vegetation. These experiments are obviously an improvement on using paper strips but they clearly also have problems in their interpretation, using small fragments of plant material (needed to fit in their apparatus) is obviously not the same as using intact forest vegetation. Obviously, burning entire forests is a difficult experiment to do and effectively impossible at oxygen levels different from the current 21%. However, it is relevant to these arguments that not all forests ignite at current levels of atmospheric oxygen, even if the vegetation is dry. During the Second World War the German Luftwaffe accidentally carried out a fascinating series of experiments by repeatedly dropping incendiary bombs on all types of British vegetation under most conceivable weather conditions, and the deciduous woodland of England did not burn even when very dry (Edlin, 1966). This suggests that 21% may be significantly below the level where uncontrollable fires are found in all types of wet vegetation. These new experiments by Wildman and colleagues (along with the old experiments by the Luftwaffe?) suggest that the 25% upper limit of Watson *et al.* (1978) may be too low. Other approaches to estimating the oxygen content of the atmosphere over Phanerozoic times also suggest periods of more than 25% oxygen (see chapter 6 of Berner 2004 for a review) and there seems to be an increasing agreement that levels of at least 30% are not ruled out by the continual presence of terrestrial vegetation (e.g. Bergman *et al.*, 2004). However, one possible complication is that these discussions tend to assume that the nitrogen levels of the atmosphere have not significantly changed, this may well be correct; however, since the mixing ration of oxygen and nitrogen is relevant in determining flammability any change in nitrogen would be important (Lovelock, 2006).

High oxygen levels may be associated with the ecology of the coal-forming forests. During the Permian and Carboniferous (approximately 250–360 million years ago) large coal deposits were formed, probably owing to a shortage of lignin degrading species (Section 3.1). As described above the amount of oxygen entering the atmosphere by photosynthesis is largely controlled by the burial of organic material, so an increased level of oxygen would be expected at this time (Robinson, 1990), indeed this is the geological period when several lines of evidence suggest oxygen levels above 25% (Berner, 2004). C_4 photosynthesis was very rare or absent until about 7 million years ago (Retallack, 2001), so the plants of the Carboniferous are all likely to be using the standard C_3 pathway. This suffers from photorespiration, the competition between O_2 and CO_2 for sites on the enzyme rubisco (ribulose-1,5 bisphosphate carboxylase-oxygenase). The relative rates of photosynthesis and photorespiration lead to the fractionation of carbon isotopes so that (Berner, 2004):

$$\Delta^{13}C = a + (b-a)\, C_i/C_a$$

where C_i = intercellular CO_2 concentration, C_a = CO_2 concentration of the air external to the leaf, $a = \Delta^{13}C$ which occurs through diffusion through the stomata, $b = \Delta^{13}C$ for the fixation by rubisco.

If O_2 increases then more CO_2 will be out-competed for access to active sites on rubisco, resulting in a higher intercellular concentration in the leaf (C_i), from the equation above it can be seen that this should lead to an increase in $\Delta^{13}C$. Experimental work by Beerling *et al.* (2002b) showed that raising O_2 to 35% increased $\Delta^{13}C$ relative to plants grown at 21% O_2. The same authors also looked at $\Delta^{13}C$ in fossil plant material and found evidence of high O_2 levels during the Permo-Carboniferous, as predicted by the higher burial of terrestrial organic matter at this time.

Such high levels of oxygen have many implications for the ecology of the Earth system, beyond the increased risk of large forest fires. For example, higher levels of O_2 could have caused a greater production of stratospheric ozone, therefore creating a better ultra-violet shield (Beerling *et al.*, 2002). In addition very large insects—such as dragonflies—are well known as fossils from this time. For most invertebrates, size is partly controlled by the ability to gain enough oxygen by passive diffusion which will be easier at higher partial pressures of oxygen—ρO_2 (Berner, 2004). Such invertebrate gigantisms are also seen from marine habitats, for example the Permo-Carboniferous period was also inhabited by several genera of large eurypterids (Sea Scorpions)—one of which was until recently miss-identified as a giant spider (Selden *et al.*, 2005), and for good reason, casts of the fossil certainly look very spiderlike to my non-specialist eye. So these large invertebrates provide at least circumstantial evidence for increased ρO_2 during this period.

Irrespective of the upper limit of oxygen (25% or 35%) it is clear that the oxygen concentration of the atmosphere has stayed between certain limits over the last 350+ million years, indeed Catling and Claire (2005) suggest that it has remained stable at $0.2 +/- 0.1$ bar over the Phanerozoic. This appears surprising since the total oxygen content of the atmosphere is cycled through the biota once approximately every 4,500 years (Lenton, 2001). If this turnover happened on a time scale of 100 million years then this stability would not be very surprising, however, with a turnover on the time scale of thousands of years there is a need for some regulating mechanism to explain the relative stability. Since the proposed regulating mechanisms (e.g. Lenton and Watson, 2000b; Lenton, 2001) involve carbon sequestration they will be discussed in the next chapter (Section 8.5).

7.5. **Is photosynthesis a fundamental process?**

Photosynthesis is obviously not necessary to life. Non-light-based ecosystems are known on Earth from sites such as around hydrothermal vents and some apparently sealed caves; however, many of the organisms involved in these systems use oxygen which is the product of oxygenic photosynthesis on other parts of the planet (Santini and Galleni, 2001). Oxygenic photosynthesis appears necessary for multicellular organisms, long food chains and the high biomass seen on Earth. As such it is potentially very important in allowing life to have large planet-wide effects; however, it is not necessary for this, as the proposed Archean CH_4-dominated ecosystem would have greatly modified Earth's greenhouse effect (Box 7.2). It is possible that oxygenic photosynthesis is of such importance as a long-term source of energy to a planet's ecosystem that it could evolve on most planets with persistent ecologies. Indeed, Fenchel *et al.* (1998, p. 272) have speculated that it is very likely so that 'oxygen-rich atmosphere may be an expected outcome for any terrestrial planet with an abundance of water, illuminated between about 300 and 800 nm, and characteristics otherwise suited for organic life'. However molecular studies have led to the general conclusion that, on Earth, oxygenic photosynthesis evolved only once (work reviewed by Kühlbrandt, 2001 and Nisbet and Fowler, 2003). There are two ways of viewing these molecular results: one is that oxygenic photosynthesis is very unlikely and that on Earth we were lucky, it evolved, but only once because the necessary merging of photosystems I and II is a very unlikely event. Alternatively, one could take the view that it is not that unlikely but that once it has evolved on a planet there is no vacant niche for oxygenic photosynthesis, so it cannot evolve again because of the presence of competitors. An informal engineering analogy would be to point out that once the motor car had been invented it made much more sense to modify existing designs than invent the whole idea again from scratch—which would be too expensive to do with competitors already producing working vehicles. These two arguments have all the usual difficulties associated with discussions of the origin of life, as we only have one example it's difficult to say much about probabilities with any confidence!

Why is oxygen so important for a lively biosphere? It is an oxidant, indeed it is exceeded in its reduction potential by only fluorine, chlorine, and a few other chemical 'species'. Importantly, it is relatively stable under the conditions relevant to biology, which allows living systems to have control over its behaviour in chemical reactions (Fenchel *et al.*, 1998). In addition, oxygen is common in the Universe, for example it is the third commonest element by mass in our solar system comprising 0.91% of the total (Gribbin, 2000).

There is a huge difficulty with answering the question 'is oxygenic photosynthesis so likely to arise on a planet that it should be considered a fundamental ecological process?' As pointed out above the difficulty is the same as we face in answering questions about the probability of life evolving on a planet, we only know one example—the Earth—and have no idea if our planet is typical or extremely unusual! However, with oxygenic photosynthesis there is some hope that future editions of this book could be revised in the light of real data. We do not need to visit a planet to find oxygenic photosynthesis (although a visit would be necessary to find *small* amounts of microbial life) as we can potentially detect oxygen-rich atmospheres remotely from Earth (Lovelock, 1965; Sagan *et al.*, 1993; Schindler and Kasting, 2000). I am hopeful that this will be technologically possible within the next few decades, hence the possibility of these data being available for some future edition of this book.

Whereas currently it is difficult to say anything definite about the likelihood of photosynthesis (or indeed life!) on other planets, it is possible to theoretically investigate the conditions which could lead to photosynthesis and the production of oxygen in quantities large enough to be detected by Earth-based astronomers. Wolstencroft and Raven (2002) attempted to model the factors which may limit—or allow—oxygenic photosynthesis on Earth-like planets. Their model suggests different levels of photosynthetic O_2 generation, depending on the type of star the planet was orbiting. This could aid the selection of Earth-like planets for detailed spectroscopic study once the technology is available.

7.6. Overview

Photosynthesis—both anoxygenic and oxygenic—allows access to new sources of energy on a planet. Oxygenic photosynthesis has the potential to create an oxygen-rich atmosphere and so allow aerobic respiration which yields much higher amounts of energy than anaerobic respiration (fermentation) and so allows longer food chains and more complex ecosystems. The amount of oxygen added to the atmosphere is intimately linked to the burial of organic matter in sediments and contrary to the impression given in many ecology textbooks marine phytoplankton are crucially important in maintaining the levels of atmospheric oxygen on Earth.

Anoxygenic photosynthesis will have a positive Gaian effect by providing a new and important source of energy, thereby increasing the probability of the long-term survival of life on the planet. Oxygenic photosynthesis is more problematical: it could potentially have either a positive or a negative effect. As with anoxygenic photosynthesis it provides an important energy source but the oxygen given off is likely to be toxic to organisms evolved in anoxic conditions (Fenchel and Finlay, 1995); could such 'oxygen pollution' kill off all life on a planet (perhaps by destroying some crucial guild of anaerobic microbes or, in the case of the early Earth, destroying the methane greenhouse)? If so, oxygenic photosynthesis has the possibility of behaving in a negatively Gaian manner. However, it is possible that these ecological problems associated with the creation of an oxygen-rich atmosphere may be less serious than is sometimes assumed. Clearly the photosynthetic organisms themselves must have evolved tolerance to oxygen; the same is presumably true for organisms which lived in close proximity to the oxygen producers (e.g. feeding on the surface of photosynthetic microbial mats). The genetic variation required to deal with oxygen-rich conditions are likely to have been widely available long before global oxygen levels started to rise (Wilkinson, 2003). In addition, the Earth still has lots of species of anaerobic microrganisms, they are just restricted to oxygen-free habitats such as some waterlogged sediments or the guts of many animals (Fenchel and Finlay, 1995).

At the current state of our knowledge it is impossible to know whether we should expect most biospheres to evolve oxygenic photosynthesis. However, improvements in telescope technology should allow us to look for oxygen-rich atmospheres around distant Earth-like planets and help answer this question over the next few decades.

8. *Carbon sequestration*

8.1. Carbon sequestration and landscape change in northwest England

During the eighteenth century in much of Europe there were significant changes in the approach to knowledge, as well as moral and social values, which slowly start to appear more modern in their outlook. Between 1724 and 1727 Daniel Defoe—best known today as the author of *Robinson Crusoe*—published his *Tour thro' the whole Island of Great Britain* describing a journey around the country in which he praised the enterprise, commerce and industry of Eighteenth century Britain (Porter, 2000). Chat Moss, a 2,450 ha expanse of peat bog between Liverpool and Manchester, did not meet with his approval. He wrote that 'The surface at a distance, looks black and dirty, and indeed frightful to think of, for it will bear neither horse or man, unless in an exceedingly dry season, and then not so as to be passable . . . what nature meant by such a useless production 'tis hard to imagine; but the land is entirely waste except . . . for the poor cottagers fuel [i.e. the use of dried peat as a fuel source], and the quantity used for that is very small'.

Chat Moss today is very different (Fig. 8.1). Drainage of these boglands started a few decades after Defoe was writing, and in the early 1760s the Duke of Bridgewater extended his recently constructed canal onto the northern part of the bog, both to help drain the land and to allow spoil from his coal mines to be dumped there. By the end of the eighteenth century parts of the most eastern area of the bog had been leased with the aim of draining it and turning it into agricultural land. This campaign of reclamation escalated when the Liverpool to Manchester railway opened in 1830, crossing the bog—a considerable engineering feat because of the very wet and unstable ground conditions—so making access for reclamation much easier (Wilkinson and Davis, 2005). Today most of the area is drained and used for agriculture and commercial peat extraction, with a small area currently managed for nature conservation. The depth of peat has greatly reduced, with an estimated 0.5–1.0 m lost during the past 50 years (Hall *et al.*, 1995). As peat is largely composed of the incompletely decomposed

Fig. 8.1:   Part of Chat Moss, once a large area of peat bogs between Manchester and Liverpool in northwest England; it has now been reclaimed for agriculture and other uses. This has led to the breakdown of the peat and the release of CO_2 back into the atmosphere. Note the drainage ditch in the foreground and the bank, with trees, in the distance which marks the line of the Manchester to Liverpool railway line. This is now raised above the surrounding landscape because of the shrinkage of the peat—both through its breakdown by microbial respiration and the physical loss of large quantities of water from the system.

remains of plants, and so is effectively temporarily sequestered carbon, this loss of peat will have returned CO_2 fixed by photosynthesis over thousands of years from the peat to the modern atmosphere.

This example of peatlands illustrates the potential for life to sequester carbon in sediments, something which I will argue is fundamental to any Earth-type ecological system and is a process of great practical importance in the context of 'global warming' on Earth. One of the objectives of this chapter is to give a longer time scale Earth Systems context to this crucial problem, a context which is usually missing from the standard ecology texts.

8.2. A tale of two cycles: the short- and long-term carbon cycles

Look in most university-level ecology textbooks and there is likely to be a diagram labelled 'The carbon cycle', and as most of these books are organized around a hierarchical series of entities, this planet scale system will probably be towards the back of the book. Although often described as *the* carbon cycle such diagrams are usually depicting only a small part of the planet's carbon cycle, often referred to as the 'short-term carbon cycle' by those with a more geological perspective on these things (e.g. Berner, 2004; Kump *et al.*, 2004). These ecology texts usually either miss out the two-way fluxes of carbon between the rocks and the surficial system (i.e. oceans, atmosphere, life, and soils), or restrict it to considering the burning of fossil fuels. This is a reasonable approximation on the time scale of thousands of years; however, this is a tiny fraction of the time span of life on Earth. To put the role of life in its proper Earth Systems context we need to consider the 'long-term carbon cycle'. A generalized mass balance equation for the long-term cycle can be written as (Berner, 2004)

$$dM_c/dt = F_{wc} + F_{wg} + F_{mc} + F_{mg} - F_{bc} - F_{bg}$$

See Table 8.1 for a definition of these terms and Fig. 8.2 for a graphical representation of these fluxes. The main fluxes in this equation which are described in most ecology texts are the movement of organic carbon into (F_{bg}) and out of (F_{wg}) organic sediments, along with the release of CO_2 from volcanoes (F_{mc}). In the case of the peatland described in the introduction to this chapter the key factors were peat formation (a short-term version of F_{bg}) and the breakdown of peat sediments after drainage (F_{wg}).

Two key factors in the long term, or geological, carbon cycle are the fluxes of carbonate carbon (F_{wc} and F_{bc}), these are often intimately connected with the weathering of silicate minerals and are usually overlooked in descriptions of the carbon cycle in ecology textbooks. The main processes in the long-term carbon cycle can be summarized in two chemical equations (Berner, 1998; Watson, 1999). The first of these is the familiar equation for oxygenic photosynthesis (or respiration if read from right to left) described in the previous chapter

$$CO_2 + H_2O \leftrightarrow (CH_2O) + O_2$$

Table 8.1: Important variables in the long-term carbon cycle (following Berner, 2004).

M_c	= Mass of carbon in surficial system
F_{wc}	= Carbon flux from weathering of Ca and Mg carbonates
F_{wg}	= Carbon flux from weathering sedimentary organic matter
F_{mc}	= Degassing from volcanism, metamorphism, and diagenesis of carbonates
F_{mg}	= Degassing from volcanism, metamorphism, and diagenesis of organic matter
F_{bc}	= Burial flux of carbonate carbon in sediments
F_{bg}	= Burial flux of organic carbon in sediments
F_{wsi}	= Carbon flux from weathering Ca and Mg silicates due to transfer of atmospheric carbon to Ca and Mg carbonates

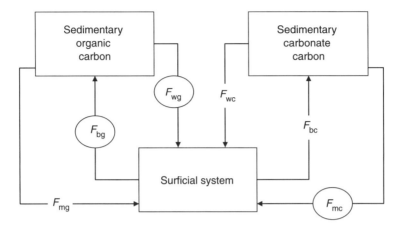

Fig. 8.2: Schematic representation of the long-term carbon cycle (after Berner 1998, 2004). See Table 8.1 for description of the various fluxes, those usually considered in ecology textbooks are encircled.

In this context it should be read as an equation for 'net photosynthesis', that is photosynthesis minus respiration, which is the amount of organic matter buried in sediments. The second key equation is that for the weathering of silicate minerals namely

$$XSiO_3 + CO_2 \leftrightarrow XCO_3 + SiO_2$$

where 'X' stands for either Ca or Mg. Read from left to right this equation symbolizes the uptake of CO_2 from the atmosphere during the weathering of silicate minerals and its transfer to HCO_3^-; which is washed into the oceans where it is deposited as $CaCO_3$ or $MgCO_3$. Read from right to left the equation summarizes the breakdown of these carbonate minerals at depth within the Earth and the subsequent release of CO_2 back to the atmosphere (Berner, 1998, 2004).

The carbon storage capacity of the sediments is far greater than that of the surficial system, as such on a time scale of millions of years the flux of carbon between these two systems must be in a steady state (Berner, 2004). So that

$$dM_c/dt = 0$$

Even a minor departure from this would, over geological time, lead to colossal changes in the CO_2 content of the atmosphere. In terms of the silicate weathering system described above it must be the case that (Berner, 2004)

$$F_{wsi} = F_{bc} - F_{wc}$$

In the context of this book it is obviously of interest to ask what, if anything, is the role of ecology in the long-term stability of this system?

8.3. **The role of life: from geochemical cycles to biogeochemical cycles**

Feedback loops leading to regulation do not necessarily require life as part of the system, as is demonstrated by the example of the natural nuclear reactors at Oklo (Section 6.5). In the case of silicate weathering Walker *et al.*, (1981) pointed out a potential abiotic negative feedback mechanism which could stabilize the Earth's surface temperature over geological time. They suggested that if the partial pressure of carbon dioxide (ρCO_2) in the atmosphere increased this would raise the surface temperature of the planet, which would both directly increase the rate of silicate weathering (as most chemical reactions increase in rate with temperature) and indirectly increase weathering through greater evaporation from the oceans leading to increased rainfall. Increased silicate weathering would, over geological time, reduce the atmospheric levels of CO_2. A decrease in CO_2 reduces the surface temperature through a reduced greenhouse effect, so there is a mechanism for the regulation of surface temperature on Earth which would apply even if life had never evolved (or arrived) on the planet. It is important to note that regulation is a term with many assumptions behind it, which has led to some confusion in the Earth Systems science literature (Box 8.1): however, to an ecologist it clearly implies 'control according to defined rules or principles' (Calow, 1999). This definition certainly includes the Walker *et al.*'s (1981) mechanism as an example of regulation, even though the Earth's temperature has not been static over geological time (Schwartzman, 1999; Kump *et al.*, 2004).

Walker *et al.*'s (1981) model was purely abiotic; however, Lovelock and Watson (1982) quickly pointed out that terrestrial life was probably crucial to a full understanding of the effects of silicate weathering on atmospheric CO_2 and climate. In the absence of terrestrial life it is likely that soil would be rare due to erosion by water and wind action. With the first terrestrial 'vegetation', such as a stabilizing crust of bacteria, algae, and lichens (see Fig. 1.2), soils would be able to start to resist erosion (Schwartzman and Volk, 1989). This simple story is complicated by the existence of a number of Archean palaeosols, dating back to some 3,500 million years ago (Retallack, 2001): the general assumption being that this is long before any life on land (but see Schwartzman (1999) for a dissenting view). Clearly the—much later—evolution of vascular plants will have allowed much deeper soils to develop.

Box 8.1: **What is regulation?**

The regulation of population size through density-dependent processes (Section 3.4) has been a major topic in population ecology; however, it has historically proved remarkably difficult to unambiguously demonstrate this process in wild populations (Sinclair, 1989). One of the most straightforward approaches is to experimentally alter the size of a population and to see if it tends to return to its pre-perturbation size. However, this is often impracticable and we are left with the statistical analysis of time series data to search for regulation.

There are similar problems in the ecology of the Earth System (e.g. Volk, 2003). A crucial question is what would we expect to see if oxygen or carbon dioxide was regulated over geological time? Some people clearly expect that a regulated system should stay the same from time period to time period, so for example oxygen content of the atmosphere should stay constant over geological time. However, we know from population ecology that regulated processes can show a much more complicated behaviour. Consider the difference equation analog of the differential logistic equation, key to much of theoretical population ecology.

$$N_{t+1} = N_t \exp[r(1-N_t/K)]$$

where N is population size (at time t or $t+1$), r is the population growth rate, and K the carrying capacity. Classic studies by Robert May (1974) showed that by changing the value of 'r' a diversity of outcomes were possible. These ranged from a stable population to a cyclical one or even chaotic behaviour; although this equation clearly has regulation around 'K' built into it.

A more intuitive feel for these problems can be gained by thinking about heating systems regulated by a thermostat, such as the heating/air conditioning system in a building or a hot water shower. Ideally you would want the temperature of the air or water to stay at exactly whatever temperature the thermostat had been set to. However, stand under an old hot water shower and you may feel the water getting hotter, then colder, then hotter again as the water temperature fluctuates around the set point; the thermostat is still regulating the temperature although it's not staying constant with time (the record for the Earth's surface temperature through the glaciations of the past 2 million years show an intriguingly similar pattern).

An additional problem with the idea of regulation is that merely saying that a system is 'regulated' misses out important information about how the

---

Box 8.1: *(Continued)*

'thermostat' is set. Consider a house, if you are told that the heating is well regulated then this gives you important information about conditions in the house—you would normally expect that the temperature will not change much over time irrespective of what happens to the temperature outside. However, some important information is missing, you would not know whether you needed to wear lots of warm clothes in the house or whether you could comfortably walk around naked unless you have been told what temperature the thermostat is set at. One of the great strengths—and provocations—of Gaia theory is that it not only suggests large-scale regulation involving life, but specifies that this will be for a 'habitable state for whatever is the current biota' (Lovelock, 2003, p. 769).

This box may appear to be a long digression from the topic of this chapter, however, my experiences in discussing 'regulation' in an Earth Systems' context over the past few years suggests that it is a concept which is prone to much confusion.

Biology increases the rate of silicate weathering by a number of mechanisms (Lovelock and Watson, 1982; Schwartzman and Volk, 1989; Berner, 2004). For example, the respiration of soil-living microbes increases ρCO_2 in the soil and the plants and their associated mycorrhizae produce a range of organic acids. In addition, a soil stabilized by life provides a much greater surface area over which chemical weathering can take place; Schwartzman (1999, p. 51) calculated that a hypothetical soil 1m deep composed of particles which were cubes of 1 mm^2 (for ease of calculation) has 6,000 times the potential reactive surface of an impermeable bare rock surface.

These arguments suggest that the first terrestrial 'vegetation' could have had important effects on the rate of silicate weathering and hence carbon sequestration. It therefore seems interesting to ask, when was the first life on land? Just as with the origin of life on Earth or the age of the first eukaryotes, this turns out to be a difficult question to answer. Traditionally, palaeontologists have thought that land plants evolved during the Devonian with little terrestrial vegetation before this time. However, it now seems likely, mainly based on evidence from fossil spores, that there was some terrestrial vegetation during the Ordovician (Kenrick and Crane, 1997; Retallack, 2000). The extent of this vegetation is currently unclear, with many authors still suggesting a late Silurian/early Devonian timing for *extensive* terrestrial vegetation (e.g. Kenrick and Crane 1997; Beerling *et al.*, 2001; Beerling, 2005). Prior to the colonization of the land by true plants, the general assumption is that there would have been terrestrial microbial crusts since the late Proterozoic (Kenrich and Crane, 1997); however, such life may be much older, for example in an important textbook on palaeosols Retallack (2001, p. 200) concluded that 'the idea of life on land as far back as 3500 [million years] is no longer outrageous speculation . . . indeed there is isotopic and geochemical evidence for a variety of microbial ecosystems in Precambrian paleosols'.

Whereas the history of the early terrestrial 'vegetation' is uncertain, by the mid-Devonian there is clear evidence for forest soils (Retallack, 1997, 2001). The arguments of Lovelock and Watson (1982) suggested that such well-developed soils should have a potentially important effect on atmospheric CO_2 through their effects on silicate weathering. At the end of the 1980s Tyler Volk and David Schwartzman (Volk, 1987; Schwartzman and Volk, 1989) formally modelled these effects, showing that they were indeed important, the size of the effect on atmospheric CO_2 depending on the exact values used to quantify the biotic enhancement of weathering. Schwartzman and Volk (1989) calculated that if this biotic enhancement of weathering was 10, 100, or 1,000 times the abiotic rate then the current temperature of an abotic Earth would be either 15°C, 30°C, or 45°C warmer than currently observed. As they pointed out the two higher figures would tend to give a current surface temperature above that which can be tolerated by eukaryotes (see Table 1.1). It seems that on Earth the biotic enhancement of weathering may have been crucial to the continued survival of eukaryotes. As the sun's energy has increased (Sagan and Chyba, 1997), the CO_2 greenhouse has decreased helping to regulate the Earth's surface temperature. However, recent work by David Beerling, Robert Berner, and colleagues (described in the next section) has complicated this interpretation by suggesting a range of new *positive* feedback loops involving plants and CO_2.

Life also plays a profound role in the global carbonate cycle through the influence of marine plankton. On a geological time scale, marine sediments can be a major sink for carbon in carbonate form (either as calcium or magnesium carbonate). The role of life in this system became particularly important during the Mesozoic with the evolution of calcareous microplankton along with a general diversification of planktonic communities (Rigby and Milsom, 2000). The precipitation of carbonate is described by the following summary equation (Ridgwell and Zeebe, 2005):

$$Ca^{2+} + 2HCO_3^- \rightarrow CaCO_3 + CO_2 + H_2O$$

This process is surprisingly counter intuitive in its effects on carbon sequestration. On the time scale of interest to most geologists it causes a removal of carbon from the oceans and atmosphere, by the addition of carbon to marine sediments (F_{bc} in Fig. 8.2). However, on the much shorter time scales which usually dominate ecological research there is an increase in CO_2 in the ocean and atmosphere (Ridgwell and Zeebe, 2005).

8.4. The co-evolution of plants and CO_2 on Earth

Carbon dioxide is a key raw material for photosynthesis, as such it seems likely that changes in the atmospheric concentration will affect plants. As I have described above, the presence of land plants also had a big effect on CO_2 concentration, so we have a fascinating feedback system between plants and CO_2. When plants moved onto land water loss would have become more of a problem—which is obviously not the case for aquatic plants. In response to this they evolved a cuticle which is relatively impervious to water. However, photosynthesis requires that CO_2 can enter the plant so this cuticle has to be pierced by stomata, which allows air to enter the plant but also water loss from the plant (Givnish, 1987; Grace, 1997). As such, there is a tradeoff between conserving water and accessing CO_2. From this it follows that the lower the CO_2 content of the air the greater the terrestrial plant's potential problem with water loss, as it will need more stomata to access CO_2. For example, preserved leaves of the dwarf willow *Salix herbacea* from sediment samples from the last glacial/interglacial cycle, show stomatal densities increasing during low CO_2 glacial periods (Beerling *et al.*, 1993).

One of the interesting aspects of the palaeobotanical record is that there is a gap of at least 40 or 50 million years between the origin of vascular land plants and the first large leaves (Kenrich and Crane, 1997; Beerling, 2005). A plausible explanation for this links CO_2 levels, stomatal density, and the heat budgets of plants (Beerling *et al.*, 2001; Beerling 2005). CO_2 fell during the later Palaeozoic as terrestrial vegetation accelerated the role of silicate weathering and enhanced carbon burial (described above in Section 8.3). This reduction in CO_2 caused an increase in stomata and so increased transpiration, having the effect of allowing plants to keep large leaves cool—although the initial selection pressure was for access to declining CO_2. Today plants can on occasion maintain leaf temperatures below air temperature due to the cooling effect of transpiration (Monteith, 1981). To illustrate this consider these data collected on a sunny day in May in southwest England when the air temperature was 15°C; while grey paper in direct sunlight was 30°C black slate roof slates were 54°C; however, sunlit leaves of several species of trees were only 14°C such is the cooling power of transpiration (James Lovelock, pers. comm.). If the leaves had been at temperatures around 40°C then photosynthesis would have stopped and at the temperature of the roof slates they would have been killed (see fig. 3 of Beerling *et al.*, 2001). The cooling effect of transpiration is even more strongly illustrated by the dark red leaves of some trees in autumn. There has been a long-standing idea that such leaves may be able to maintain higher temperatures than green leaves, in the cool of autumn, because these darker pigments absorb more solar energy (Wheldale, 1916; Wilkinson *et al.*, 2002). Recent detailed measurements in New England, USA, suggest that these dark colours have no effect on temperature (Lee *et al.*, 2003), presumably because of the effect of transpiration.

Modelling studies suggest that a plant with large leaves, but the low stomatal densities of early land plants, would have been at risk from death from over-heating (Beerling *et al.*, 2001). As such it can be argued that plants responses to low CO_2 (evolving higher stomatal densities) allowed large leaves to develop (Beerling, 2005); however, this raises the question of why could not increased stomatal density evolve as an adaptation for regulating the plant's temperature, even at high ρCO_2?

One of the interesting implications of these ideas of David Beerling and colleagues is that many of the feedbacks between terrestrial vegetation and CO_2 over geological time may be positive, and so potentially destabilizing (Beerling and Berner, 2005). For example, the development of larger land plants is made possible by lower CO_2 levels and the presence of this vegetation further reduces CO_2 in the atmosphere. It is currently very difficult to quantify the overall effect of terrestrial vegetation in this context and to quantitatively compare the relative effects of negative feedback processes (Lovelock and Watson, 1982; Schwartzman and Volk, 1989) with the positive feedback processes (Beerling and Berner, 2005). Certainly Beerling and Berner (2005, p. 1302) are correct in highlighting the fact that 'insight into these critical feedbacks is, however, extremely limited'. One thing is clear, the ecology of terrestrial vegetation is crucial to a full understanding of these problems.

8.5. Oxygen and carbon sequestration on Earth

As described in the previous chapter (Section 7.4), the paired summary equations of oxygenic photosynthesis and aerobic respiration illustrate the close links between the carbon and oxygen cycles in the Earth system. They are so important that I repeat them here

$$CO_2 + 2H_2O \rightarrow (CH_2O) + O_2 + H_2O \text{ photosynthesis}$$

$$(CH_2O) + O_2 \rightarrow CO_2 + H_2O \text{ respiration}$$

Biochemically aerobic respiration is the reverse of photosynthesis, so for oxygen produced by photosynthesis to persist in the atmosphere some organic material must escape respiration by other organisms and become buried in sediments. Without such carbon sequestration oxygen cannot build up in the atmosphere.

The continuous charcoal record suggests limited variation in the oxygen levels over the past 350 million years (Lenton and Watson, 2000b; Section 7.4 of this book). The link between the carbon and oxygen cycles means that aspects of the carbon cycle have been suggested to play important roles in the regulation of atmospheric oxygen. For example, the rate of silicate weathering under the influence of vegetation and CO_2 has important implications for ρO_2. This is because increased weathering causes more phosphate—from rocks—to enter the oceans, so allowing more planktonic photosynthesis which produces a net increase in atmospheric oxygen if some of the planktonic carbon is sequestered in marine sediments. At the same time high ρO_2 increases the magnitude of forest fires, providing a negative feedback to the silicate weathering cycle by reducing terrestrial vegetation (Lenton and Watson, 2000b; Lenton, 2001; Bergman *et al.*, 2004). The long-term sulphur cycle is also linked to the carbon and oxygen cycles, for example by the oxidative weathering of pyrite FeS_2 (Berner, 2004)

$$15O_2 + 4FeS_2 + 8H_2O \rightarrow 2Fe_2O_3 + 8SO_4^{-2} + 16H^+$$

The tendency for ecology textbooks to often consider these cycles as isolated systems, which can be described independently, is obviously a gross simplification at these more geological time scales. However, the central role of life in these long-term cycles makes them of great interest to ecologists.

8.6. **Humans and carbon sequestration**

The preceding discussion highlights the role of life in carbon cycling, currently one species *Homo sapiens* is having unprecedented effects on this system. This is not a book on applied ecology, however, it is impossible to write a chapter on carbon sequestration without any mention of the effects of humans on the future of the Earth System (see Lovelock (2006) for an impassioned Gaia perspective on the perils of human-driven climate change). With 'global warming' we are not taking part in any new processes within the Earth system, but we are greatly altering the rate of some of the existing processes (see Box 8.2 for a brief history of human intervention in the Earth's carbon cycle). In the context of Fig. 8.2 we have greatly increased the rate of 'F_{wg}' by burning fossil fuels. In addition, we have altered the distribution of carbon within the 'surficial system' box by processes such as biomass destruction—which converts carbon from organic matter into atmospheric CO_2.

Box 8.2: **Chronology of events relating to humans and the carbon cycle (based, in part, on the appendix of Grace, 2004)**

Before 1800. For thousands of years human activity, such as forest clearance, has probably caused changes in the level of greenhouse gasses (e.g. CO_2 and CH_4) in the atmosphere. The climatologist William Ruddiman (2003, 2005) has estimated that we started to have effects on CO_2 levels some 8,000 years ago and methane levels some 5,000 years ago; however, these conclusions are somewhat controversial (Crutzen and Steffen, 2003).

1780–1820. Industrial Revolution—a dramatic increase in the use of coal and so a rise in atmospheric CO_2.

1859. Oil wells sunk in Pennsylvania, USA—oil is soon being produced in many countries. Also in this year the Irish scientist and mountaineer John Tyndall discovers that water and CO_2 absorb specific wavelengths of infra-red radiation and suggests that this is important in the regulation of the Earth's temperature (the 'Greenhouse effect').

1896. Swedish chemist Svante Arrhenius suggests that increasing emissions of CO_2 will lead to global warming.

1903. Ford motor company founded in USA, this quickly leads to the start of the mass production of cars.

1958. Charles Keeling starts measuring atmospheric CO_2 on Mauna Loa in Hawaii using an infra-red analyser, these measurements are still being made today (Heimann, 2005).

1968. First photographs of the Earth from deep space, these have a dramatic effect on how people view the Earth; it looks a lot smaller and easy to damage when seen from space. The astronomer Fred Hoyle (1950) predicted the psychological effect of such photos on our view of the planet almost 20 years before they were taken!

1979. James Lovelock publishes his first book on '*Gaia*', suggesting that life may be crucial to the regulation of the Earth's climate and chemistry.

1987. Ice core data from Antarctica show a close correlation between CO_2 and temperature over the past 100,000 years.

1988. Intergovernmental Panel on Climate Change (IPCC) is established.

Box 8.2: *(Continued)*

1997. Kyoto Protocol, international agreement to limit greenhouse gas emissions.

1998. The warmest year of the century, and probably of the millennium.

2001. President Bush announces that USA will not sign the Kyoto protocol.

2005. Kyoto Protocol formally came into force in February, but without the USA. However, some US states are taking their own action to try and meet Kyoto targets.

Although the silicate weathering feedbacks control ρCO_2 on geological time scales, it has little effect on time scales less than 100,000 years (Ridgwell and Zeebe, 2005)—so even though it will eventually help to reduce CO_2 artificially increased by human activity it will not do so on a time scale considered useful by most people! On shorter time scales the marine carbonate cycle does play a role in atmospheric CO_2 and this is currently a cause for concern, as increased atmospheric CO_2 causes more CO_2 to dissolve in the ocean. This lowers the ocean's pH and decreases CO_3^{2-}—one of the building blocks of calcium carbonate—according to the following equation (Gattuso and Buddemeier, 2000):

$$CO_2 + CO_3^{2-} + H_2O \rightarrow 2HCO_3$$

This can cause decreased calcification of plankton (Riebesell *et al.*, 2000). It is currently uncertain how a decrease in calcified plankton will feedback to ρCO_2 in the atmosphere. One possibility is that a reduction of carbonate production in the surface ocean would reduce the amount of CO_2 in the water therefore allowing it to absorb more atmospheric CO_2, and so act as a negative feedback on global warming (as carbonate production releases CO_2 into the water, as described at the end of Section 8.3). However, it is also plausible that carbonate production helps remove particulate organic carbon from the surface waters to ocean sediments, so less carbonate production could reduce the removal of organic carbon to ocean sediments and so act as a positive feedback on global warming (Ridgwell and Zeebe, 2005). It would obviously be good to know which of these two processes is the most important!

Probably the first direct human intervention in the short-term carbon cycle was by forest clearance. For example, 5,000 years ago the dominant vegetation type over most of Britain was various sorts of forest: ranging from Pine *Pinus sylvestris* and Birch *Betula* spp dominated forests in the north to Lime *Tilia* spp dominated forest in the southeast of England and Oak *Quercus* spp dominated ones over much of southwest England (Bennett, 1989). As agriculture developed most of this forest was lost, a trend that has been repeated in many parts of the world. Agriculture first appears approximately 10,000 years ago at the eastern end of the Mediterranean; although just as with the ages of the first life or first eukaryotes there are disagreements about the details of exact timing (Lev-Yadum *et al.*, 2000). Over the past 10,000 years the idea of agriculture has been invented on a minimum of five occasions around the world—and possibly on nine or more independent occasions (Diamond, 1997 and 2002). William Ruddiman (2003, 2005) has suggested that deforestation associated with agriculture was affecting global CO_2 levels some 8,000 years ago and CH_4 production (presumably from rice paddies etc.) rises about 5,000 years ago. As is well known there are still concerns about forest clearance in areas such as the Amazon basin.

Since the removal of forests has been a source of CO_2 replacement of forests could act as a CO_2 sink (But see Box 7.2 for possible implications for CH_4 production). For example, Rowantree and Nowak (1991) estimated that the urban trees of the USA sequestered approximately 6.5 million tonnes of carbon per year. Provided rainfall levels are still adequate forests can cover a denuded landscape remarkably quickly (Fig. 8.3). The classic example of this are the forests of eastern USA and Canada much of which were cleared during the eighteenth and nineteenth centuries to make room for European-style agriculture. However, during the mid-nineteenth century, as people moved from the farms to seek their fortunes in the newly expanding cities, much of this land went out of agricultural production and returned to forest (Foster, 1992). Although the North American example is the most well studied, giving rise to many detailed studies of 'old field succession' described in the ecology textbooks, similar things have happened in many other places. For example, between 1960 and 1980 large areas of forest were cleared in the Chorotega region of Costa Rica— this area includes the Santa Rosa National Park well known to ecologists through decades of groundbreaking work by Daniel Janzen and colleagues. Much of the forest was cleared for beef production and when, in the 1980s, beef prices fell large areas were allowed to revert to forest (Arroyo-Mora *et al.*, 2005). Both the north American and the Costa Rican examples are of areas which have returned to forest after earlier human-caused deforestation. However some areas have never had forest because they were too isolated for trees to arrive and have not had enough time for the tree growth-habit to evolve from scratch, a fascinating example being Ascension Island in the tropical south Atlantic. Much of the island is very arid, however, on Green Mountain there is much higher rainfall and considerable occult precipitation. Through human introductions of a large

(a)

(b)

Fig. 8.3: Examples of new or re-grown forests. (a) Regenerating forest in southern Ontario, Canada. This is typical of much of eastern USA and Canada in having been largely cleared of forest for agriculture in the eighteenth and nineteenth century, with later forest re-growth after many people moved to the developing sites of eastern north American during the later nineteenth and early twentieth centuries. (b) The summit of Green Mountain on Ascension Island. Prior to human

number of plant species, the mountain has gone from a treeless system in the early nineteenth century to one which today is best described as tropical cloud forest (Wilkinson, 2004a).

The archetypal pristine forest system which has become an icon for much of the environmental movement are the tropical rainforests. The examples of both dry and moist forests from Costa Rica and the Ascension Island cloud forest suggest that tropical forests may be able to recover quickly from serious human disturbance, in a similar way to the forests of eastern north America. Palaeoecological studies are now starting to show that many areas currently covered by tropical rainforest have had significant disturbance in the past. A good example of this is the Darian area of Panama, rich in endemic species, it gives the impression of untouched tropical wilderness. However, Bush and Colinvaux (1994) used pollen and other remains from sediment cores to show a long history of human disturbance. Their data suggested significant forest clearance for agriculture which was abandoned after the Spanish conquest, so this apparently primeval tropical forest was largely 350 years old. Similar evidence for widespread human disturbance of apparently 'virgin' rainforest has recently been described from Africa and Asia, as well as South and Central America (reviewed by Willis *et al.*, 2004); clearly rainforest systems may sometimes be much more resilient than has previously been assumed. It is interesting to compare this rapid recovery of recent forests with the evidence for rapid recovery of terrestrial biomass following the supposed impact event at the end of the Cretaceous (Section 6.4).

Probably the most important factor in this fast recovery, and its associated carbon sequestration, is adequate rainfall. In the Ascension Island example, forest was able to colonize Green Mountain because of the trade winds blowing large amounts of moisture onto the summit (Ashmole and Ashmole, 2000; Wilkinson, 2004a). Modelling suggests that destruction of large areas of the Amazon forest could greatly reduce rainfall due to reduced transpiration and so lead to a climate unsuitable for tree growth (Betts, 2004). Therefore a large biomass of vegetation in the Amazon is crucial to the continual existence of a climate able to support a large biomass of vegetation. Presumably the clearances of tropical forest now known from the archaeological record were not extreme enough to have such wider climatic effects.

---

intervention the largest common plants here were ferns (although the ferns visible in the photograph are introduced species). Now the summit is cloud forest dominated by Bamboos with trees covering much of the rest of the mountain. The sunny conditions in the photograph are deceptive, for much of the time the summit is enveloped in cloud which provides the water supply for this lush growth of vegetation on an otherwise arid island.

One important result of thinking about the carbon cycle from a more geological perspective is that it draws attention to the crucial role of soil. In the long-term carbon cycle this soil is important because of its effects on silicate weathering, in the shorter cycle it is the amount of organic matter contained in the soil which is important; how this will change with global warming is currently a large and important unknown in the behaviour of the Earth System (Cox *et al.*, 2000; Lenton and Huntingford, 2003). The importance of soils in carbon sequestration is illustrated by a survey of Monks Wood, a well-studied deciduous woodland in southeast England, where it was estimated that approximately 129 tha^{-1} of carbon was contained in the vegetation (including roots) and 335 tha^{-1} in the soil (Patenaude *et al.*, 2003).

The extent to which land use changes have affected soil carbon content is not well quantified but it is potentially very important, for example some cultivated soils are known to have lost over half their soil organic carbon (Lal, 2004). Therefore the management of soils, as well as above-ground biomass, is important in the context of increasing atmospheric CO_2 and climate change. The effect of past human land use on soils is also a potential complication for the ideas of Ruddiman (2003, 2005) on an early start to human effects on the global carbon cycle and climate. Many of the past modifications of the land will have reduced soil organic matter which is consistent with Ruddiman's hypothesis; however, there are important exceptions. For example, in Britain clearance of forest from wet, often upland, parts of the country triggered the formation of blanket bog (Moore, 1975, 1993), such systems can sequester more carbon than the previous forest. However, the anaerobic nature of peat bogs creates an additional complication as microbes within them produce CH_4 (a greenhouse gas) as well as sequestering carbon. Peatlands are important in the global short-term carbon cycle as they contain around 455 Gt of carbon, which equates to the amount of carbon in all living organisms on the surface of the Earth (Moore, 2002). As such destruction of peatlands—be they the temperate bogs described in this chapter's introduction or tropical peatlands such as those in southeast Asia— is clearly important for the CO_2 content of the atmosphere.

8.7. Carbon sequestration as a fundamental process

For any planet with carbon-based life, carbon sequestration appears unavoidable. Firstly some organic matter is going to escape consumption by other organisms and become entombed in sediments. On any planet with oxygenic photosynthesis this is particularly important as it potentially leads to increasing oxygen levels in the atmosphere. On Earth the silicate weathering cycle is a key process in the long-term carbon cycle, this is likely to be true of any Earth-like planet; as such the evolution of terrestrial life with its effects on weathering rates is likely to increase rates of carbon sequestration. The obvious conclusion is that biologically controlled carbon-sequestration is likely to be present on any planet with carbon-based life. However, it is important to realize that in general this aspect of life is an unavoidable by-product, not a specific adaptation. Natural selection is likely to favour individuals who utilize carbon sources, to ones that allow organic matter to escape to sediments. However, in the case of silicate weathering, adaptations which allowed access to mineral nutrients in rocks will have the additional effect of increasing weathering rates (Schwartzman and Volk, 1989).

Currently on Earth our industrialized societies play a major role in the carbon cycle and if Ruddiman (2005) is correct so have many pre-industrial societies. Since nothing can be said with confidence about the probability of life on other planets, it is clearly impossible to say much about the likelihood of species with human-like levels of agriculture and industry, although many have attempted educated guesses (e.g. Shklovskii and Sagan, 1977; Davies, 1995). However thinking about sources of energy for such a hypothetical species suggests that they will be closely linked to carbon sequestration on their planet. An obvious early source of energy for a species which is going to develop agriculture and industry is burning biomass—it is hard to see how an ecology without oxygenic photosynthesis would be 'lively' enough to produce such a species, so the presence of an oxygen-rich atmosphere ideal for 'burning' is a good guess. Later this biomass burning is likely to be followed by burning fossil fuels, as such an intelligent species is likely to have profound effects on its planet's carbon cycle—before perhaps turning to technologies such as nuclear fission or fusion?

If this reliance on fossil fuels during industrial development is indeed crucial then it allows interesting predictions about the history of industrial societies on any planet. As Fred Hoyle (1964) realized, this makes industrialization a 'one shot affair'. Consider a major catastrophe overcoming industrial society on Earth, resetting the historical clock to a medieval level of technology. With easily accessible 'coal gone, oil gone, high grade metallic ores gone, no species however competent can make the long climb from primitive conditions, to high-level technology' (Hoyle, 1964, p. 64). Therefore although we can say little that is sensible about the probability of an intelligent species with high technology evolving on another planet, we can predict with reasonable certainty that it will only happen once on each planet, or at least be separated in geological time by a period long enough for the regeneration of fossil fuel supplies outcropping at the planet's surface.

8.8. **Overview**

Whereas traditionally ecology textbooks only discuss the short-term carbon cycle, the role of life has been crucial in the geological long-term carbon cycle through processes such as silicate weathering. Interesting arguments have been put forward for the co-evolution of CO_2 levels and terrestrial plants—with adaptations to lower CO_2 levels allowing large leaves to evolve. It seems clear that on Earth without the effect of life (especially terrestrial vegetation) our planet would currently have a greenhouse-effect-controlled temperature which would appear to rule out the survival of eukaryotic life. As such it would appear reasonable to claim that carbon sequestration has a positive Gaian effect, however, this is probably a local conclusion, suitable for the Earth, but not able to be generalized to all other planets. If, as in the case of the Earth, the output of the planet's star has increased over time then reducing the CO_2 greenhouse may help to maintain life-friendly conditions, if the planet is not too far from the star. As the sun is thought to be a fairly typical 'main sequence' star (Narlikar, 1999), this conclusion will apply to many planets. However, if the planet is towards the outer limit of the habitable zone in its solar system (as defined by temperature affecting the availability of water) then lowering the greenhouse effect could lead to the cooling of the planet and reduce the life span of its biosphere. As such the fundamental ecological process of carbon sequestration could have a positive or negative Gaian effect depending on the particular situation on a given planet. More generally these ideas illustrate the importance of biomass as a key feature of global ecologies (see Chapter 6), the effects of vegetation (or plankton) on carbon cycles is more directly linked to available biomass than species richness.

Part III Emerging systems

There is a continued circulation of the matter of the surface of the globe from lifelessness to life, and from life back again to lifelessness.

Huxley (1887, p. 228)

9. *Nutrient cycling as an emergent property*

9.1. **The paradox of the Goldfish**

The Goldfish *Carassius auratus* is native to Asia and parts of Eastern Europe (Lever, 1977; Maitland and Campbell, 1992). It has a long history in captivity having been kept in China since at least the time of the early Sung Dynasty (AD 960–1279), and in Britain the first written record is from 1742 when Thomas Gray wrote his *Ode on the death of a favourite cat, drowned in a tub of Gold Fishes* (Lever, 1977). In Britain it became traditional for children to keep Goldfish as pets, in the past these were often kept in a glass bowl apparently devoid of other life forms—but in practice the water would have been rich in microbes. A hypothetical microbe-free Goldfish bowl forms a useful counter-thought experiment to the approach I have taken in this book. There is only one species, the fish, which survives because of a reliable input of fish food from outside the system. Many of the candidate fundamental processes described in Part II of this book are absent from this Goldfish bowl system. A more realistic view of this thought experiment would be to imagine a planet inhabited by a single species, such as a microbe living in its ocean and being fed by a slow but constant input of extraterrestrial organic matter.

The approach I have taken in this book causes me to argue that the power of natural selection, and the important idea of ecological tradeoffs, would lead to the evolution of a more diverse biota with multiple guilds utilizing each other's waste products; hence producing many of my fundamental processes and leading to the emergence of nutrient cycling. Until other planets with life are discovered, it seems impossible to test these ideas with real data. However, this does not necessarily mean that these ideas have to be based on mere intrinsic plausibility, as it is possible to approach these questions in a more rigorous manner with both 'evolution in a test tube' studies and computer simulations.

9.2. **Tradeoffs and *in vitro* evolution**

In Chapter 4 I argued for a role for tradeoffs in generating the conditions needed for biodiversity, suggesting that this idea should be given much more prominence in ecology than is currently the case. Although often not explicitly emphasized this idea has been implicit in some ecological studies for many decades, a good example of this being the classic studies by Joseph Connell (1961) on barnacle distribution on the sea shore at Millport in Scotland. His results, described in many text books, implied that the fast growing barnacle species *Balanus balanoides* appeared to trade off its ability to survive the stresses of life high on the sea shore against its ability to grow quickly, whereas the slow growing *Chthamalus stellatus* could survive in such stressful conditions. Philip Grime (1974, 2001) has similarly suggested a tradeoff between rapid growth and an ability to cope with stressful conditions in plants. These ideas have been around for some time but their general importance as an organizing principle in ecology has tended to be overlooked.

Studying evolution in action by experimental methods is often difficult because of the time scales involved. Quickly evolving systems, such as bacterial populations evolving *in vitro* or virtual organisms evolving *in silico* can produce results in more realistic timescales. A particularly interesting study, relevant to this book's principal thought experiment, was carried out by Rainey and Travisano (1998); using the common aerobic bacterium *Pseudomonas fluorescens*, which readily exhibits evolution under *in vitro* conditions. In their experiments they provided the bacteria with a range of ecological opportunities—afforded by spatial structure in the microcosm tubes. In a matter of days, replicate populations developed very similar morphological diversity, unless spatial structure was removed by shaking the tubes. The types of spatial structure available to the microbes is very similar to that available in a Goldfish bowl—namely glass sides, surface film, and open water. Tradeoffs appear very important in the evolution of this diversity, for example adaptations which suit colonization of the surface film are not good for colonizing the open water. As it is hard to envisage a planetary surface devoid of spatial structure these experiments provide some support for the idea that evolution would quickly lead to the breakdown of a single-species Goldfish-bowl biosphere.

The experiments of Rainey and Travisano (1998) also illustrated some very important points about frequency-dependent changes in organism's fitness, which are very relevant to this book's main thesis. One of the important arguments against seeing a role for life in global regulation is the question, 'What about Ghenghis Khan species?' that is, species which effectively take over and destroy any community they reach (Pimm, 1991, p. 190; see also Hamilton, 1995). The *Pseudomonas* microcosm experiments illustrate one possible answer to this question as these bacteria showed frequency dependence in the fitness advantages of various genotypes, so that they had advantages when rare which disappeared when they were common. For example, one of the *P. fluorescens* morphs was named 'wrinkly-spreader' by Rainey and Travisano (1998), it formed a layer on the surface film of the water which gave it access both to oxygen in the air and nutrients from the microbial medium. However, as it becomes very common the wrinkly-spreader (which was reducing access to oxygen for the other morphs) became a victim of its own success, as the surface film mats increased in weight a point was reached where they could no longer support themselves on the surface and they sank. It is possible that such Ghengis Khan species may often have the seeds of their own destruction hidden in their initial destructive successes.

These *in vitro* systems using *P. fluorenscens*, or similar microbes, are a very informative way of studying evolutionary ecology—although there are obvious questions about how well the results scale up to the far larger spatial and temporal scales of planetary ecology. They have been used to investigate a range of processes as well as the effects of the environment on adaptive radiations, as described above. For example, they have been used to study the role of parasites in host diversification (Buckling and Rainey, 2002), and the effects of disturbance and productivity on diversity (Buckling *et al.*, 2000; Kassen *et al.*, 2000, *et al.*, 2004). Such systems can potentially provide evolutionary insights into several of the fundamental processes described in this book.

9.3. **The emergence of biogeochemical cycles**

Clearly the development of tradeoffs and adaptive radiations are highly relevant to the emergence of biogeochemical cycles, and the evolution of microbes *in vitro* is a very useful way of studying such things. The other main approach currently available is evolution *in silico* as exemplified by artificial life studies of the evolution of nutrient cycling, especially the work of Keith Downing and colleagues (Downing and Zvirinsky, 1999; Downing, 2000, 2002, 2004). These two approaches are important as other potentially powerful methods are unavailable—for example, while we only have one planet with life comparative approaches are unavailable, although historically they have been very important methods in ecology and evolutionary biology (Bradshaw, 1987).

The GUILD model of Downing and Zvirinsky (1999) is particularly informative, it simulates the evolutionary emergence of a web of interacting species whose combined effects can potentially control aspects of their chemical environment. This artificial life system uses a genetic algorithm (GA) approach. GAs are computer programmes first developed during the mid-1960s which simulate the evolution of a population of genotypes, represented by a string of binary digits, with both sexual recombination and mutations of their genetic code (Holland, 1992). In the GUILD system the environment is represented by '*n*' different chemicals, while an organism's genes determine both the chemicals it feeds on and those it produces as waste products (organisms are not allowed to both feed on and excrete the same chemical). The organisms reproduce by division with genetic variation coming from both mutations and gene swaps between organisms (sex *sensu lato*). Through the production and consumption of chemicals these organisms can create local ratios of nutrients which differ from the global ratios in the simulation. In these simulations the competition for resources leads to a diversity of guilds and the whole ensemble of guilds leads to the emergence of life-sustaining nutrient cycling.

Keith Downing (2002, 2004) has also produced a second artificial life system with somewhat more realistic chemistry, which he called METAMIC. Although GUILD essentially allowed all possible chemical reactions for the same metabolic cost METAMIC used abstract chemistries which constrained the system to a smaller set of legal reactions and had more realistic relationships between energy, entropy, and biomass. The reduced number of legal reactions in METAMIC restrict the number of metabolic options for the virtual organisms. This leads to a wider range of possible outcomes for the whole system than in GUILD. The GUILD system showed a strong tendency to settle down to a stable state with regulated nutrient cycling. With METAMIC this happened in some runs of the simulation although on other occasions regulation broke down. As Downing (2004, p. 278) commented 'if METAMIC has captured some essence of the real world, then one of the basic take-back-to-the-wild lessons from the computer runs is simply that Gaian homeostasis is neither tautological or impossible, but clearly contingent on a host of physicochemical, and quite possibly historical, factors'. Planets with the correct contingent histories may be very likely to develop the types of persistent ecologies which are the subject of this book's principle thought experiment.

9.4. Cycling ratios and biotic plunder

The results of the *in vitro* and *in silico* studies described above suggest that we should expect biogeochemical cycles to appear as emergent properties of the presence of life on a planet, and that in many cases these cycles should have a homeostatic character that will tend to increase the probability of life surviving on a planet. Clearly it is these systems with some element of regulation that we would expect to survive for a long period of time and so potentially to dominate the examples of living planets in the Universe. This assumption may turn out to be wrong if life evolves from non-life very easily, which could lead to a large number of transient systems of short life expectancy which could outnumber the more long lived stable 'Gaias'. However, if it is the case that the initial origin of life is almost impossibly improbable then Earth may be the only example of such a system in the Universe. If so, the observation that Earth-type life dominates the set of biologies in the Universe is trivial since the set only has one member and so it tells us little about the probabilities of Gaias developing, only about the unlikely nature of life itself.

Assuming a Universe of relatively stable Gaias, what can we predict about the structure of their emergent cycles? A useful way of thinking about nutrient cycling is the concept of cycling ratios. This idea is due to Tyler Volk, however so far it has only been published in one of his books aimed at non-specialist readers (Volk, 1998)—although he intends to publish a more formal scientific paper in due course (Volk pers. comm.). The idea is simple, but usefully quantifies the way life on Earth cycles different nutrients. Consider calcium, life uses approximately 620 million tons a year whereas about 550 million tons is lost each year from living systems (Volk, 1998). The ratio of these two numbers being $620/550 = 1.1$; so around 10% of calcium that enters living systems is recycled once before being lost to the abiotic world. Contrast this with phosphorous where Volk (1989) suggests that the comparable figures are $230/5 = 46$. The comparison of these cycling ratios 1.1 for Ca and 46 for P illustrates in a quantitative manner that phosphorous is cycled much more often through living systems before being lost to the non-living world. Evolution of nutrient cycles, of the type seen in Downing's simulations (described above), should lead to higher cycling ratios for nutrients in short supply. In Downing's simulations his GUILD model produced cycling ratios of 30–40 while the (more realistic?) METAMIC simulation gave ones in the range 5–15. The results for phosphorous described above show that ratios of 46 are possible in the real world, even if METAMIC tended to evolve lower ratios (Downing, 2004).

The well known capacity for unconstrained biological populations to grow exponentially, suggests some potentially important generalizations about the concentration of nutrients in the environment. If all nutrients and other resources are common then this should lead to unconstrained growth of the organisms which will eventually use up the nutrients and so lead to an environment in which the nutrients are in short supply: a process which has been called 'biotic plunder' by Tyrrell (2004). It is interesting that many cases of human-caused pollution involve problems created by artificially raised nutrient levels—such as the eutrophication of water bodies in agricultural areas (e.g. Moss, 2001). Tyrell (2004) mainly bases his arguments on plankton production in ocean systems; however, his logic has much in common with some of the arguments which have been put forward to explain 'why are parts of the world green?' in the terrestrial ecological literature (e.g. Polis, 1999). Viewed from an aeroplane most oceans, and large lakes, look blue while viewed from underwater—with snorkel or SCUBA gear—visibility is often quite good. Both of these observations are true because of the low numbers of phytoplankton in most waters. This is mainly due to nutrient limitation, which can be demonstrated experimentally by adding additional nutrients. For example, iron fertilization experiments in the Southern Ocean caused the sea to turn green and the responsible microbes to sequester increased amounts of CO_2 (Abrahams *et al.*, 2000; Watson *et al.*, 2000). So the presence of life in the oceans may regulate nutrients around low levels by rapidly using up any increased nutrients.

Although Tyrell (2004) developed his ideas on biotic plunder based mainly on marine data he did suggest that a similar situation may exist in many terrestrial systems. A good example of this comes from Vitousek's (2004) long-term studies of the ecology of Hawaii. He suggests that there is a plant/soil–microbial positive feedback system which tends to lower nutrient levels over time. He suggests that high-nutrient-use efficiency by plants causes them to have lower nutrient levels in their tissues and so lower levels in the plant litter, this leads to slower rates of decomposition and hence low-nutrient availability which encourages high-nutrient-use efficiency in the plants.

How does all this fit with a Gaian view of planetary ecologies? Early versions of the Gaia hypothesis (e.g. Lovelock and Margulis, 1974) discussed a planetary 'homeostasis by and for the biosphere'. Such phraseology tends to suggest regulation for optimum conditions for life. Indeed Kleidon (2002) has recently argued that life optimizes conditions for net primary productivity, based on models of terrestrial systems rather than aquatic ones. Such definitions look unlikely to be correct in the context of the idea of biotic plunder, described above. However, current versions of Gaia stress the 'habitable state' of the planet. For example, in an essay in *Nature* James Lovelock (2003, p. 769) defined Gaia as the process by which 'organisms and their material environment evolve as a single coupled system, from which emerges the sustained self-regulation of climate and chemistry at a habitable state for whatever is the current biota'. The idea of Gaian effect, used in this book (and in Wilkinson, 2003), comes from this emphasis on maintenance of habitability which is potentially very different from the maintenance of optimum conditions. An interesting analogy with physiology is the well-known phenomena that reduced calorie intake increases longevity in many animal species (Kirkwood and Austad, 2000), clearly starvation is not optimal but it can maximize longevity in a similar way to biotic plunder creating stable regulated conditions, although ones which are not optimum for biomass production. As Tyrell (2004) pointed out the idea of biotic plunder is compatible with the idea of life being involved in regulation and that this regulation is for a habitable state. However, this habitable state is not optimal for life if your definition of 'optimum' is based on primary production or biomass, but it could be regulation for 'improved conditions' if this is defined as increasing the longevity of life on a planet—that is the Gaian effect. This appears sensible, as one can see in general terms how the presence of life as an important forcing factor in a system should make it more likely that the system stays within life-friendly conditions—although good quantitative theory is still lacking—it is not obvious why regulation for optimum conditions should be expected.

9.5. **Overview**

Arguments based on both *in vitro* and *in silico* models suggest that biogeochemical cycles will readily evolve on planets with life, along with many of the putative fundamental processes described in this book. Artificial life models illustrate the potential for these emergent cycling systems to have a positive Gaian effect. The well-known potential for exponential growth in unconstrained ecological systems suggests that these emergent systems will often regulate their environments around low-nutrient states (biotic plunder) rather than at states which optimize productivity. This provides a context in which to understand why human-caused eutrophication is often a problem, if the natural state of most systems tends towards nutrient scarcity then it is not surprising that raising nutrient levels dramatically can sometimes have unwelcome effects. In the context of biotic plunder it makes sense to define Gaia in relation to prolonged habitability of a planet but not as a process which maximizes biological productivity at any one point in time.

10. *Historical contingency and the development of planetary ecosystems*

10.1. Carus and the thunder bolt: chance and change in history

In AD 283 the Roman Emperor Carus was struck by lightning and killed, within a couple of years of this fatal accident both of his sons had been murdered and power had passed to a new Emperor whose approach to the administration of his empire was very different from that of Carus (de la Bédoyère, 2002). A single unpredictable accident had changed the history of the most powerful group of humans on the planet.

The narrative of human history is full of such pivotal chance events, for example the seas surrounding Britain have regularly illustrated the important role of accident in history. In 55 BC and 54 BC the Roman army made incursions into southern England from continental Europe, and on both occasions their logistics were thrown into confusion by storms in the English channel; indeed it was not until AD 43 that the armies of the Emperor Claudius successfully invaded much of Britain (Schama, 2000). In the summer of AD 1588 there was an even better illustration of the power of historical contingency in these same waters as a Spanish fleet made ready to attack England. The English ships under the command of Sir Francis Drake famously attacked this armada off the south coast of England, but in truth Drake's efforts did only minimal damage to the Spanish who sailed on to meet up with reinforcements in the Netherlands. Then, in Schama's (2000) felicitous phrase, 'the wind turned Protestant' and the resulting storm caused far more damage to the Catholic Armada than Drake had been able to achieve. A different weather pattern could have led to a very different outcome for England from 1588 onwards. Given the crucial global role this country played during the eighteenth and nineteenth centuries, this one storm may have affected the histories of countries from Canada to India, and beyond.

10.2. **Historical contingency and ecology**

The idea that similar historical accidents are important in the history of life on Earth, as well as in the history of human societies, has been developed by several authors. In recent years, the most eloquent advocate of its importance on geological time scales was the late Stephen Jay Gould (1990, 2002). He developed his arguments in a geological context based, in part, on the fossil faunas of the Burgess Shale which are described in the next section of this chapter.

An interesting ecological example of the apparent role of chance is the history of the shrub *Rhododendron ponticum* in Britain and Ireland. *Rhododendron* is a large genus with much of its species richness concentrated in southeast Asia. *R. ponticum* is an evergreen shrub about 2–8 m high and is native to parts of Spain and Portugal and also the area around the Black Sea and the eastern Mediterranean (Cross, 1975). The shrub was introduced into Britain as an ornamental garden plant about 1763 and was widely planted as cover for game birds during the nineteenth century. Today it is often considered a 'major alien environmental weed' (Dehnen-Schmutz *et al.*, 2004). As Chambers (1995) has pointed out, this history makes *R. ponticum* a particularly interesting example of the apparent role of historical contingency in ecology. Its natural distribution during the Holocene suggests a species limited to southern Europe and the Middle East—the obvious guess would be that its absence further north was due to the climate. However, the human introductions show that this shrub can thrive in much more northerly climes, being found in northern Scotland and at altitudes up to 600 m in the mountains of northwest England (Preston *et al.*, 2002). Obviously climate is not the key factor in the limited natural range of this species; this conclusion is supported by fossils of both pollen and seeds, from sediments from a previous interglacial in Ireland—however no fossil remains are known from mainland Britain (Godwin, 1975; Chambers, 1995). Although this interglacial was probably warmer than the Holocene it still illustrates the shrub growing at much higher latitudes than its recent 'natural' distribution.

Over the past 10 years there has been a growing realization that rare long-distance dispersal events (e.g. seeds carried by birds or unusual weather conditions) are often crucial in determining the rate of expansion of plant ranges (e.g. Wilkinson, 1997; Cain *et al.*, 1998; Higgins *et al.*, 2003), although as so often in ecology and evolutionary biology Darwin (1859) had got the basic ideas correct in the nineteenth century, but had been ignored by most twentieth-century ecologists. Indeed, when I wrote my own papers on this topic in the late 1990s (Wilkinson 1997, 1999b), Darwin's work provided some of the few published relevant experimental studies I could find on this topic! Throughout geological history organisms have had to alter their ranges in response to climate change—a behaviour which has been particularly necessary during the glacial/interglacial cycles of the Quaternary (Huntley and Webb, 1989; Bennett, 1997). The most plausible explanation for the distributional history of *R. ponticum* is that, by chance, in at least one previous interglacial its seeds were carried to Ireland which it successfully colonized; but that a lack of such chance events during the Holocene has kept it out of Britain and Ireland until the eighteenth-century human introductions. However, the climate and other conditions are obviously suitable and have enabled it to become a successful species when introduced by human action. If plants track climatic change with the aid of rare accidental events then this clearly raises questions about the availability of such events over the short time scales predicted for human-caused 'global warming'. In addition the habitat fragmentation caused by humans breaking up semi-natural vegetation with urban or intensively agricultural areas is likely to make it harder for plants to move through many current landscapes (e.g. the models in Dyer, 1995).

Additional examples of the role of chance in ecology are provided by models of the interface between population ecology and biogeography (O'Regan *et al.*, 2002; Wilkinson and O'Regan, 2003). There is a general consensus in conservation biology that animals which are restricted to areas of a limited size (e.g. an island or a nature reserve) will only survive in the long term if this area is of a certain size—for a well-known example see Newmark (1987; but see Parks and Harcourt (2002) for an alternative, and probably more realistic reinterpretation). One of the difficulties with biogeographical studies is that no two cases are exactly the same, so that if you are working with species which are too large for artificially controlled experiments to be feasible, then every example is a special case. Consider a large organism, such as a big cat, living on an isolated area such as an island. Using a mixture of modern and historical data it is possible to conclude that a given population of that species is able to survive on an island of a given size (or alternatively we could conclude that the island is too small, if the species is known from fossils or historical sources but became extinct at a time when humans were probably not responsible). The obvious question of relevance to this chapter is, would such a population always survive or always go extinct on this island, or is an element of chance involved? One approach to answering this question would be to try and find other islands of the same size which have had populations of this species and ascertain if they managed to survive. However, there are bound to be other factors which vary between these islands (different climates, different competitor, or prey species etc.), so they are not really good replicates of the situation on the first island.

A theoreticians approach to this problem is to run replicate stochastic population models on a computer, so that you can be certain that each replicate does indeed have identical starting conditions—although as models they will necessarily be simplified caricatures of reality. The results of modelling of this type is shown in Fig. 10.1. In these VORTEX models, a population of tigers

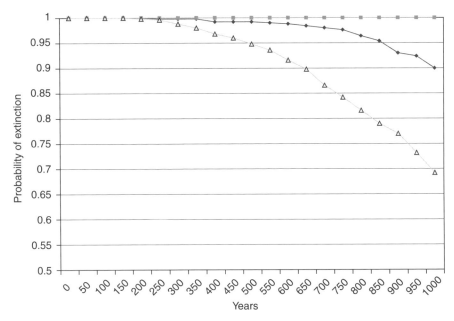

Fig. 10.1: VORTEX models illustrating the role of chance in the survival of big cat populations. The squares (■) show results for replicate tiger *Panthera tigris* populations on an island the size of Bali, whereas the diamonds (♦) show results for leopards *P. pardus* on an identical island. The triangular symbols (△) show the results for a cat of a similar size to the extinct *P. gombaszoegensis*, mainly based on the similar-sized jaguar *P. onca*, isolated in a glacial refugium the size of Italy (for background details on the palaeobiology of *P. gombaszoegensis* see O'Regan and Turner, 2004). These models were run using VORTEX version 8, with each curve being based on 500 iterations; for full details of the tiger and leopard models see Wilkinson and O'Regan (2003); and for *P. gombaszoegensis* see O'Regan et al. (2002). The key point illustrated by these results is the element of chance in the survival of two out of the three cat populations modelled. This arises directly out of the population dynamics in these models, as no environmental catastrophes or other additional variables (such as human hunting) were included. A ceiling-carrying capacity, without an explicit description of density dependence, was used in all three models (cf. the models in Fig. 3.5; the legend to this figure also includes other details and caveats about VORTEX models). Lacy (2000) provides a general description of VORTEX.

Panthera tigris on an island the size of Bali always survived for a period in excess of 1,000 years, while 10% of the leopard *P. pardus* populations on this virtual island became extinct. Fig. 10.1 also shows the results for a population of medium-sized cats isolated in an area the size of Italy, and largely based on the population biology of the modern jaguar *P. onca*. This was an attempt to study the ecology of the now extinct 'European Jaguar' *P. gombaszoegensis* and its ability to survive glacial periods in southern European refugia. This is a point of some interest as many temperate European organisms may have spent approximately 90% of the Quaternary in such limited ranges (refugia), as the glacial periods lasted much longer than the interglacials (see Bradshaw, 1999, fig. 2). Although this putative Italian refugium was large (and most of the assumptions in the model were conservative, tending to increase the chance of survival) approximately 30% of the replicate populations became extinct over a 1,000-year period. These models illustrate the potentially large role for chance processes in the survival of populations of larger organisms. The situation may be very different for many free-living microbes where huge population sizes may make extinction very unlikely (Fenchel, 2005), so chance may play a much smaller role in the ecology of the microbes which dominate many Earth System processes.

10.3. **Historical contingency and the Earth System**

Moving out of the Quaternary to the full expanse of geological time, there is huge scope for historical contingency. The most spectacular examples are the impacts of large meteorites on Earth (e.g. Grady *et al.*, 1998). These can potentially cause extinctions ranging from the loss of an endemic species of limited range size living in the area of impact to the destruction of all life on Earth, depending on the size of the meteorite. Other potential triggers for catastrophic global extinction events include Snowball Earth or the oxygenation of the atmosphere. The survival of such mass extinctions may often be something of a lottery, with chance playing a role in which lineages survive to take part in subsequent adaptive radiations (Gould, 2002).

Probably the most well-known argument for the importance of historical contingency in palaeobiology is that associated with the Cambrian faunas recorded in the Burgess Shale. In part its prominence in the literature is due to Gould's (1990) accessible account of these faunas and what he viewed as their theoretical implications. The Cambrian is marked by an apparent 'explosion' in the diversity of the fossil record due to the evolution of shells and other hard body parts which more readily form fossils than does soft tissue. However, many of the animals in the sea at this time were still soft bodied and so less likely to fossilize. For a full picture of Cambrian biodiversity we need sites with unusual conditions (usually anoxic sediments) which allow soft tissue to fossilize. The most famous of these is the Burgess Shale in Canada (Conway Morris, 1989, 1998; Gould, 1990) although many other sites are now known from around the world (e.g. Conway Morris *et al.*, 1987; Zhang *et al.*, 2001). These Burgess Shale type faunas are characterized by great morphological diversity, although the exact extent of this is controversial (Conway Morris, 1998). During the 1980s it was suggested that there was probably a large element of chance in which of these lineages survived to give rise to the animal phyla we know today (Conway

Morris, 1989; Gould, 1990), although Conway Morris (1998) is now less happy with this highly contingent interpretation.

A variety of explanations have been put forward to explain the 'explosion' in fossil diversity during the Cambrian. These range from suggestions that this explosion is mainly an artefact of better fossil preservation to ecological mechanisms, such as a greater role for predators, or explanations based in developmental biology and genomic change (Gould, 1990; Conway Morris, 2000). In the context of this book some of the more interesting explanations are the ones based on Earth System processes. For example Lenton and Watson (2004) suggested that an early colonization of the land by microorganisms increased the rate of silicate weathering (Section 8.2) which, via effects on atmospheric O_2 concentration, triggered the ecological changes behind the Cambrian explosion. A more conventional ecological explanation is the suggestion that the evolution of hard parts was driven by predation (e.g. Conway Morris, 2000). In this context the identification of diffraction gratings on the surface of some Burgess Shale animals fossils (Parker, 1998) is very interesting, as colour now plays a prominent role in the ecology of predation, both in crypsis and warning colouration (Ruxton *et al.*, 2004). Several of these mechanisms may have contributed to the real explanation, as it is often the case in ecology that there are several mechanisms involved and the key question is not which mechanism is correct but, what is the relative contribution of the different mechanisms (Lawton, 1996)? However, if Gould (1990) was correct, the main message of these faunas is one about the importance of historical contingency.

10.4. **Overview**

The ecology of a planet is influenced by historical processes, at any stage in its development the current conditions of life on a planet form the starting point from which new conditions develop. This means that over time an ever-increasing number of historical accidents will be incorporated into the system and so the role of past history will become increasingly important. This happens across a range of scales, from the chance of long-distance dispersal of *Rhododendron* seeds to the survival of mass extinction events.

The idea of historical contingency is a simple one, hence the relative brevity of this chapter, and yet it is crucially important in understanding much of ecology. Consider oxygenic photosynthesis and the rise of oxygen levels in the atmosphere, which has rightly been described as a 'defining moment in Earth history' by Canfield (1999; see also Lenton *et al.*, 2004). This constrained the possible subsequent trajectories of ecological development on Earth, for example in relegating anaerobes to limited oxygen free habitats. Such historical contingencies mean that the fundamental processes described in this book are necessary, but not sufficient, to explain the history of a biosphere; this is similar to the relationship between the ideas of *evolution* and *natural selection*. The opening sentence of the preface to one of the greatest theoretical books on natural selection states: 'Natural selection is not Evolution' (Fisher, 1930). Although natural selection is necessary to explain evolution it is not sufficient to understand how evolution has proceeded in a particular case as we need to know a range of historical details, many of which are effectively determined by chance events (Frank, 1998). In the same way that the history of England could have been different if a storm had not caught the Spanish Armada.

11. *From processes to systems*

11.1. **An Earth Systems approach to ecology**

My approach in this book has been to use an astrobiological thought experiment as a way of thinking about ecology on the scale of the whole Earth. This approach to ecology may be very important if the subject is to make major contributions to the mitigation of human-caused changes to the planetary system, as confronting problems such as global warming require the understanding of very large-scale systems, such as the atmosphere of the whole planet. There are several differences in emphasis between my approach to ecology and more conventional approaches exemplified by university-level textbooks on the subject. Firstly microorganisms feature much more prominently in this book than they do either in ecology texts or ecology journals. For example during 2005 only 5% of the papers in the international ecology journal *Oikos* had a significant microbial content, and these tended to be dominated by eukaryotes (see Wilkinson (1998) for additional analyses of the abundance of microbes from ecological journals and conferences). Other ways this book differs from conventional ecology are a tendency to give increased emphasis to the importance of biomass, rather than species richness, and an increased emphasis on parasites—rather than predators. The relative lack of interest in parasites, compared to the huge literature on predation, was also identified as a problem with traditional ecology 10 years ago by Hamilton (1995) in his essay review of Pimm's (1991) book *The balance of Nature*.

A succinct way of viewing the differences in the approach of conventional ecology and the more planetary perspective used in this book is provided by Lewontin (2000). He summarized the traditional evolutionary approach as a pair of differential equations.

$$\mathrm{d}E/\mathrm{d}t = f(E)$$

$$\mathrm{d}O/\mathrm{d}t = f(O, E)$$

where change in the environment (E) with time (t) is a function of various environmental variables—where E is described in purely abiotic terms. The evolutionary change in organisms (O) over time is a function of both the abiotic environment and the biota. Lewontin suggested that the real situation is better represented by a pair of *coupled* differential equations.

$$dE/dt = f(O, E)$$

$$dO/dt = f(O, E)$$

So that changes in the abiotic environment are a function of both E and O as are changes in the organisms. At the larger spatial and temporal scales discussed in this book these coupled differential equations are clearly a better way of thinking about the way things work, and are a return to the theoretical approach of Alfred Lotka in the 1920s (Kingsland, 1995). The key point is the two-way feedback between the biotic and abiotic elements. Consider one without the other and you are likely to be missing crucial parts of the system.

Richard Lewontin has been making similar points since at least the early 1980s. These evolutionary ideas have recently been developed in much greater detail under the title of 'niche construction' (e.g. Odling-Smee, *et al.*, 1996, 2003; Laland, *et al.*, 1999). This approach has focused on the evolutionary implications of the coupling between organism and environment, and has developed a theoretical background derived from the mathematics of population genetics. A related idea that was developed during the 1990s is that of ecological engineering (Jones *et al.*, 1994, 1997), this is a more ecological idea—compared to the evolutionary approach taken by niche construction. It points to the way in which some organisms greatly alter their environment both for themselves and other species; either by their physical presence (e.g. trees) or their behaviour (e.g. beavers or humans). Unlike niche construction, with its mathematical underpinning from population genetics, ecological engineering is grounded in natural history observation. Indeed the original paper in *Oikos* (Jones *et al.*, 1994) describing the idea is highly unusual for a late twentieth-century theoretical paper in being all natural history and no mathematics!

Both these approaches have emerged from within ecology and evolutionary biology within the past 20 years. However the approach taken in this book has been heavily influenced by Gaia, an idea developed in the late 1960s and early 1970s (see Lovelock, 2000b, for a historical account). Gaia is clearly an ecological idea, although not developed by an ecologist. Indeed this approach was largely developed in ignorance of ecology during the 1970s; although Lovelock has since come to see both A.J. Lotka and G.E. Hutchinson as forerunners of Gaia. Hutchinson was an ecologist whose work fits into the approach taken here very well, his wide academic interests included biogeochemistry and the application of cybernetic theory to ecological systems (Slobodkin and Slack, 1999). His work formed part of a mid-twentieth-century interest in processes at the ecosystem scale (see Odum, 1969, for a classic review), sometimes referred to as 'systems ecology'. Such large-scale studies seemed to go out of fashion by the 1980s and early 1990s, but are currently becoming more common again (Fig. 11.1), this revival of ecosystem scale studies is probably due to the interest and concern over global warming and related phenomena. In the context of this speculation it is interesting that Chambers and Brain (2002), in an extensive analysis of the published literature, suggested that 1988–1991 marked a major change in research publications, with an increase in papers on human-caused climate change and its potential effects. The book *Leaving Eden* by Nisbet (1991) provides an interesting example of how a talented and imaginative Earth scientist viewed 'global change' at the time of this paradigm shift: influenced by Gaia he argued for a physiological approach to understanding the Earth System, but in a much more applied context than I take in this book.

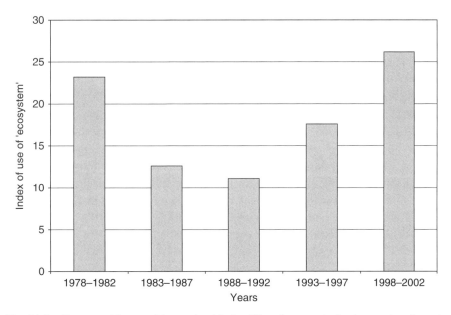

Fig. 11.1: The use of the word 'ecosystem' in the titles of papers in five top ecology journals between 1978 and 2002; data taken from appendix 1 of Nobis and Wohlgemuth (2004). The journals used were *Ecology, Journal of Animal Ecology, Journal of Ecology, Oecologia*, and *Oikos*. All these journals showed an increase in the numbers of papers published per year over this time period, to remove this effect I have divided the number of papers mentioning ecosystem in each time period by the total number of papers published in that period then multiplied by 1,000, to give an index based on whole numbers. Note the decline during the 1980s and early 1990s.

11.2. **Cybernetics, regulation, and Gaia**

Gaia not only implies strong coupling of life and the abiotic environment, as in Lewontin's coupled differential equations, but also regulation. Clearly the question of the existence of such regulation and how it might work, are important research questions. At a major international meeting on Gaia in Valencia in 2000 (for the proceedings see Schneider *et al.* (2004)), Lee Kump suggested during the discussions that Gaia theory had shifted thinking away from cataloguing fluxes and pools (e.g. of carbon in the biosphere) to asking questions about control systems and feedbacks (reported by Gillon (2000)). Such questions are crucial if we are to understand the effects of adding more CO_2 or CH_4 to the atmosphere.

The initial ideas on Gaia were developed using the analogy of the Earth as a super organism (see Section 2.2), this has worried many eminent biologists (e.g. Dawkins, 1982; Ehrlich, 1991; Williams, 1992), because planetary regulation appears to be a more difficult phenomena to explain than homeostasis in an organism. In most cases it is clearly in the interests of, for example, the cells in a multicellular organism to benefit other parts of that organism; however this is not the case with Gaia. Consider the production of DMS by some marine algae (Section 6.3). This chemical has several functions in the Earth System; for example, closing a loop in the sulphur cycle by transporting sulphur from the oceans back to the land and also for its role in cloud formation (Charlson *et al.*, 1987). However, if these functions were costly then it should pay an individual cell not to contribute DMS to the Earth System. This is the well-known theoretical problem of 'cheats' in mutualisms (e.g. Wilkinson and Sherratt, 2001; Yu, 2001), but at a planetary scale.

Such arguments probably mean that any Gaian regulatory mechanism cannot have evolved for the purpose of planetary regulation. There appears to be a growing consensus, amongst several theoreticians, that Gaia must be built of by-products (e.g. Volk, 1998; Wilkinson, 1999a, 2004; Lenton and Wilkinson, 2003). In the case described above the precursors to DMS would be produced for the cell's own needs, perhaps osmoregulation (Sunda *et al.*, 2002), and their role in cloud formation and the sulphur cycle is 'accidental'.

11.3. **Rotten apples: a conceptual model of the evolution of Gaia**

One of the big problems in trying to think about the evolution of Gaian type systems is the huge complexity of the Earth System. One approach is to try and find a simpler system which still exhibits many of the key problems presented by Gaia. A good candidate for such a system are the ideas on the rotting and spoiling of fruits, seeds, and dead animals, suggested by Daniel Janzen during the 1970s (Janzen, 1977, 1979). Provoked into thinking about microbial spoiling of food after accidentally buying a rotten avocado, Janzen considered the process from the microbes' perspective. He speculated that many of the chemical produced by microbes during the spoiling process (e.g. alcohols, aflotoxins, botulinin, etc.) were a product of adaptation by the microbes to prevent larger organisms eating their substrate. As Janzen (1977, p. 707) wrote, 'Fruits rot, seeds mold, and meat spoils because that is the way microbes compete with bigger organisms'.

There is a problem with this idea—which Janzen noted in his 1977 paper—that has many similarities to the evolutionary problems with Gaia. For these chemicals to have any effect on deterring larger organisms microbes would need to collaborate in producing them—just as the quantities of DMS needed to have cloud-producing effects requires many phytoplankton to produce it. There is an obvious evolutionary temptation for the microbes to cheat and refrain from producing these chemicals while benefiting from the effects of the chemicals produced by others. Janzen wondered if some kind of group, or kin, selection could provide the explanation. In an attempt to investigate Janzen's provocative idea I have been involved in trying to model the evolution of this system along with Tom Sherratt and Rod Bain (Sherratt *et al*., 2006).

We produced a numerical simulation model (summarized in Box 11.1), the basic structure of which can be understood by visualizing an orchard with wind-fall apples. These apples can be colonized by microbes of two types: spoilers which produce a distasteful chemical and non-spoilers which do not. The apples are consumed by larger organisms such as Thrushes (birds in the Turdidae). Our model suggested that if it was costly for the spoilers to produce these chemicals then it was impossible for them to persist even with the help of group selection; a result which would surprise few evolutionary ecologists. More interesting was what happened if spoiling was made cost free, then the spoilers could persist in the system but predation by the thrushes had no effect on their presence or spread from rarity. *Although the chemicals had the effect of reducing predation on the apples colonized by spoilers their production could not evolve or persist for that reason.* The effect of the chemicals on bird predation of the wind-fall apples was a product of chemicals which must have evolved and continued to be synthesized for other reasons. It is often too easy to assume that if something can be shown to have a useful effect (such as deterring thrushes) then this must be the reason for its evolution. Although this will often be the case at the organism level, this intuition will often be wrong at the Earth systems/Gaian level of organization.

Gaian mechanisms are likely to be similar in character to this wind-fall apple model. While they can have a regulatory effect on a planetary system they will have evolved, and persist, for other reasons. If Gaian processes are by-products

---

Box 11.1: **The spoiling model—a summary**

The numerical simulation spoiling model of Sherratt *et al.*, (2006) envisages a number of microbial resource patches (*n*), such as wind-fall fruits or cadavers. It comprises both continuous-time processes (growth rate of microbial populations and depletion of resources) with discrete-time processes (consumption of fruit and transmission of microbes between fruit, also fruit replacement).

The continuous-time microbial growth on each 'fruit' (*i*) was described by the following equations:

$$dR_i/dt = K_N N_i R_i - K_s S_i R_i$$

$$dN_i/dt = f_N R_i N_i - b N_i$$

$$dS_i/dt = f_s R_i S_i - b S_i$$

where R_i is the amount of resource, N_i and S_i are amounts of non-spoilers and spoilers, respectively, b is the microbial mortality coefficient, K_N, K_s, f_N, and f_s are the consumptive and reproductive coefficients for the spoilers (*s*) and non-spoilers (*N*). The model was run using a range of 'fruit' consumption rules (e.g. *x* fruit with highest $R_i - c S_i$ were selected: where *c* is a description of the importance of avoiding spoilers) and a variety of 'fruit' replacement rules (e.g. new fruits are infected with a ratio of $S_i : N_i$ drawn from a randomly selected 'fruit' in the population).

The results of this modelling was that if spoiling carries a cost (i.e. the microbe must use some resources to make the spoiling chemicals) then not even group selection could enable it to persist. In the absence of a cost to spoiling then the spoilers were able to persist even without the selective feeding of large animals on the 'fruit', but could spread from rarity only under a limited set of conditions.

---

then the 'cheat' problem disappears; however, there is still a problem of why such processes should be regulating if there is no mechanism for the selection of regulation. This is an area requiring much new theoretical work; although some possible explanations were outlined in Section 6.5 and by Wilkinson (2004b). Unless one takes a view based on an extreme version of the anthropic principle, there must be mechanisms which explain the apparent regulation of the Earth System over time.

11.4. **Conservation biology and the Earth System**

This is not an applied ecology book; however, it is worth briefly considering how an Earth Systems perspective affects approaches to nature conservation. As with academic ecology, conservation biology often focuses on entities such as species or habitats, indeed Jepson and Canney (2003) have suggested that there is a 'recognition by scientists that their ability to represent nature in units . . . creates the opportunity to integrate ecological theory with neoclassical economics' as an entity, such as a species, can be ascribed a value—this is something Jepson and Canny correctly consider not to be in the best interests of conservation. Species have often been used as what Simberloff (1998b) has referred to as 'umbrellas' in an attempt to conserve biological entities. The idea being that a charismatic rare species (such as a big cat) can be used to justify the preservation of an area of land which will also help preserve many other species (Simberloff, 1998b). The idea of nature reserves or parks is also entity based.

This book has focused on processes rather than entities, taking a planetary scale approach. What would this suggest for conservation? Haila (1999b) has argued that from this perspective the important things to safeguard are the processes themselves not the named entities, be they tigers or rain forests. If the processes are protected then rainforest can recover and large predators can re-evolve. In this context, the discussion in this book would suggest that we should pay more attention to biomass than to the numbers or variety of entities. Indeed important blocks of natural or semi-natural vegetation can be missing the kinds of entities that conservationists most often value, for example Simberloff (1998b) pointed out that the 6,812,000 ha Tongass National Forest in Alaska did not have a single species which qualified for protection under the United States Endangered Species Act. Yet this National Forest would be contributing ecological services, as indeed do wholly unnatural vegetation types dominated by introduced species, although such vegetations are usually considered of no conservation interest (e.g. Wilkinson, 2004a).

It is possible that the great reduction in natural and semi-natural vegetation in many parts of the world could be compromising some regulatory processes. Clearly this is an important research question. What is clear is that 'habitats' which are not normally considered of any interest by Nature conservationists, such as ones rich in introduced species or plantation forestry, may still be playing a valuable role in the functioning of the Earth System, and that *such systems may be becoming increasingly important as the more natural Systems decrease in area and biomass*. Such vegetations have traditionally not attracted the interest of most academic ecologists, it is time this changed as we need to know much more about them and be able to say with more certainty how good they are as surrogates of natural vegetation. There is some evidence that this is starting to happen,

(a)

(b)

Fig. 11.2: Examples of manmade habitats of conservation interest because of their biodiversity. Traditionally managed hay meadows are a good example of unnatural systems which are high in biodiversity. The disturbance of cutting prevents a few plant species from dominating the site, as does the relatively low-nutrient levels due to the lack of modern chemical fertilizers. Our modifications to

just before I handed the manuscript of this book over to the publisher an interesting paper authored by 18 plant ecologists was published in *Global Ecology and Biogeography* arguing that we should start thinking much harder about the ecology and management of these 'novel ecosystems' (Hobbs *et al.*, 2006). These vegetations may even be increasingly important for the conservation of entities such as rare species as climate change suggests that relying on nature reserves of reasonably pristine habitats is not a long-term solution as these reserves will come to have the 'wrong' climate for the species they were originally set up to protect (Huntley, 1994; Pitelka *et al.*, 1997). However, amongst all these problems, there are some reasons for limited optimism. Semi-natural and, in some cases, wholly artificial vegetation can sometimes rapidly develop and carry out various Earth System processes (Section 8.6) and some more low-intensity human land uses produce habitats which are greatly valued by many conservationists (Fig. 11.2). While the Earth Systems approach is currently in its infancy it may eventually lead to a very different emphasis for the conservation movement, a focus on entities may be no longer practical in many cases in our rapidly changing world; a preservation of the processes which will eventually allow recovery may be the best we can achieve. For someone who grew up as, and still is primarily a naturalist, these are not easy conclusions to reach; the great, and rapidly declining, diversity of life on Earth is of huge importance to me.

11.5. Concluding remarks

A system's view of nature, replete with feedbacks, is a remarkably recent way of viewing the world. In European thought the dominant technological metaphor of many centuries was clockwork: something that should run perfectly, without the need for feedback, once it had been set going. During the eighteenth century, especially in Britain, it became increasingly common to find ideas of 'checks and balances' and self regulation; especially in political and economic theory (Mayr, 1986; Henry, 2002). Technology utilizing feedback had been known from at least

the world do not have to cause mass extinctions, however, such low-intensity use probably requires a much lower population size than currently found on Earth, if it is to be widespread. (a) Forest Meadow at Råshult (the birth place of Linnaeus) in southern Sweden. Pollen analysis has shown that this meadow system was created, from woodland, about 900 years ago (Lindbladh and Bradshaw, 1995). It is still managed by traditional methods, with the aid of subsidies, as this is no longer an economic way to farm in Sweden. It is rich in plant species, as is appropriate for the birthplace of one of the greatest names in plant taxonomy! (b) Traditional, species rich, alpine meadow in the Italian Dolomites. Because of the steep terrain it is still managed by traditional methods (the mechanical harvester in the picture being little larger than a domestic lawn-mowing machine).

the first century AD, however these were 'toys' rather than devices of practical use. Feedback starts to become technologically important in eighteenth-century England: first for devices for regulating windmills and later for regulating steam engines—for example the famous Boulton-Watt centrifugal governor of 1788 (Mayr, 1986). These ideas of feedback and checks and balances moved from economics and technology to medicine during the nineteenth century, most importantly in the work of Claude Bernard in France on the idea which would later become known as homeostasis (Porter, 1997). Almost 100 years later, during the Second World War, physiologists studying the aiming of anti-aircraft guns came to appreciate the crucial importance of feedback processes in the interaction of muscles and brain, an insight they took back to postwar medical research (Miller, 1978). It is possible that our understanding of such complex feedback-rich systems is still in need of new concepts and approaches before we can develop good quantitative theory. For example, Joel Cohen (2004), in a recent essay on Mathematics and Biology, suggested that understanding coupled atmospheric, terrestrial, and aquatic biospheres in the context of global biogeochemical processes may require major mathematical innovations and that research in this area may encourage such developments in much the same way that attempts to study the dynamics of interacting species stimulated Alfred Lotka's mathematical ideas in the early twentieth century.

In the context of Gaia it is probably relevant that James Lovelock spent the first part of his scientific career working in medical research—for example playing a crucial role in elucidating the mechanism by which cells are damaged by freezing, a result with huge implications for reproductive medicine and the storage of eggs, sperm, and embryos (Pegg, 2002). Ideas of feedback already existed in the environmental sciences, such as the 'systems ecology' of Hutchinson and the Odum brothers during the mid-twentieth century (for an Earth Science example see the 'flow diagrams' in Holmes, 1944). However, Gaia was marked out by a particular emphasis on feedback and regulation, which was unusual in the environmental sciences in the 1970s, indeed the interest in ecosystem processes in ecology was about to suffer a decline (Fig. 11.1). As John Lawton (2001) has pointed out in an editorial on Earth Systems Science in *Science* 'James Lovelock's penetrating insights that a planet with abundant life will have an atmosphere shifted into extreme thermodynamic disequilibrium, and that the Earth is habitable because of complex linkages and feedbacks between atmosphere, oceans, land and biosphere were major stepping-stones in the emergence of this new science.' As far as I have been able to discover, the term Earth System Science was coined at NASA during the mid-1980s. At least no one working in the area who I have spoken to can remember the term used prior to this.

The Earth System is obviously very complex and we are currently altering it in many ways. A Gaian approach is a way of trying to organize a lot of information in a way that allows one to ask interesting, and hopefully useful, questions. In particular, it forces us to think hard about feedbacks and gives microbes the central place they deserve in ecology. In the context of the processes described in this book the ecologically most important group (Kingdom *sensu* Margulis and Schwartz, 1998) are the prokaryotes followed by the single-celled eukaryotes and then the fungi and plants (perhaps of equal importance). The least important group is the animals. The striking thing about this ranking is it is almost exactly opposite to the amount of attention given to these groups by ecologists!

In the terminology of Vespsäläinen and Spence (2000) Gaia is not really a scientific theory (a relatively constrained statement about the world susceptible to a single test which could find it wanting), but an 'explanatory framework'. Such frameworks are 'road maps to solutions, rather than solutions themselves . . . so that an investigator can pick and choose what is required to effectively understand a specific event or situation' (Vepsäläinen and Spence, 2000, p. 211). It is now clear to many scientists that it is impossible to understand a planet such as the Earth without considering multiple feedbacks between life and the abiotic environment. It is also clear to many that the Earth exhibits a certain amount of regulatory behaviour, in which life is intimately involved. Key questions for the future are about the strength of these regulatory processes and the mechanisms underlying them, and how the fundamental ecological processes contribute to the working of the Earth System.

Glossary

This book considers ecology in the broader context of Earth Systems Science, as such it ranges across a number of scientific disciplines. This created problems in the use of various technical terms, which come not only from ecology but also from a range of other disciplines such as astronomy, climatology, geology, chemistry, and physics. In an attempt to reduce this problem the glossary defines many of technical terms used in this book, in addition I have included some comments which would have broken the flow of my principal arguments if inserted in the main text. In composing the definitions I have assumed that anyone looking up a term probably comes from an area of science different from the one where the term is in common currency. As such the glossary attempts to give the basic idea behind the term, rather than aiming for a fully rigorous formal definition; in addition, where I cite a text for further information I have, in most cases, deliberately chosen less technical texts (usually at a level accessible to someone starting to study that subject at university). Two excellent textbooks written at this level are particularly worthy of note to help fill in gaps in scientific background; for traditional ecology, see Cotgreave and Forseth (2002) and for Earth Systems Science see, Kump *et al.* (2004). Both of these texts provide more detailed explanations of many of the terms briefly defined below. A good dictionary of Biology is Hale *et al.* (2005), written by three of my colleagues at Liverpool John Moores University; while Calow (1999) is a very useful dictionary of Ecology. The glossary is followed by a brief summary of the Geological time scale on Earth.

Abiotic environment: Traditionally, ecology textbooks split the environment into 'biotic' (the biological aspects of the environment) and the 'abiotic' (the physical aspects). Abiotic would include factors such as aspects of climate, soil pH, or oxygen concentration of the atmosphere; however, the approach taken in this book makes it clear that this classification is often unhelpful as most of the traditional abiotic aspects of the environment are affected (perhaps even regulated?) by biology.

Adaptive radiation: The evolution of a wide range of related forms from an ancestral species.

Albedo: Also called 'reflection coefficient', the reflectivity of an object. For example the albedo of fresh snow is about 0.95 whereas a flat calm ocean can be as low as 0.20 (see Pielou (2001)).

Anthropic principle: The idea that only some types of universe could possibly support observers (astronomers, etc.) therefore it is not surprising to find that we live in such a universe. This has been an important, but also difficult and controversial idea in astronomy (see Smolin (1997), for a sceptical but relatively non-technical discussion).

Archaea: Sometimes called 'Archaeobacteria'. Treated by most modern biology textbooks as a kingdom of Prokaryotes (*q.v.*), based mainly on the genes determining the small subunit ribosomal sequences (see Postgate (2000)). However I favour the currently less popular position which views them as a subdivision of the kingdom Bacteria (see Cavalier-Smith (2004), and references therein, for a technical justification of this view or Margulis and Schwartz (1998), for a more general discussion).

Archaean: The geological eon covering the period of time from 2,500 million years ago until approximately 3,800 million years ago. The start of the Archaean is not well defined, a logical place to draw the lower boundary would be at the first appearance of life, however as described in Chapter 1 the dating of this event is highly controversial.

Autotroph: An organism which can use non-organic sources of energy. The most well-known examples being green plants using solar energy.

Biomass: The total mass of living material. For example, the biomass of a forest would be the combined mass of every organism living in that habitat.

Biome: An area of the world with similar ecologies examples include 'Tropical Rainforest' or 'Temperate Deciduous Forest'.

Biosphere: This term has been used in several different ways. Some authors use it to mean the 'totality of living things residing on the Earth', others use it to mean 'the space occupied by living things' while others use it to refer to 'life and life support systems' (Huggett, 1999). In this book the second definition is used; that is, the Biosphere is that part of the planet occupied by life. The third 'systems' definition has much in common with Gaia. (*q.v*). For an English translation of the classic text on 'The Biosphere' first published in 1926, see Vernadsky (1998).

Boreal forest: The vast forests of coniferous trees forming a zone between tundra and deciduous forest in the northern hemisphere.

Carrying capacity: The maximum number of individuals that an area can support, such as the carrying capacity of a particular bird species on an island. Often represented as K in the logistic equation (*q.v.*).

Chloroplast: A structure within plant cells which contains the photosynthetic pigments. These are now generally thought to have evolved from free living photosynthetic bacteria (see Wakeford (2001)).

Cloud condensation nuclei: Substances around which water droplets condense during cloud formation.

Coccolithophorids: Planktonic protist ('protozoa') covered in plates of calcium carbonate (see Margulis and Schwartz (1998)).

Cosmic rays: Elementary particles (e.g. protons, electrons, etc.) travelling through space.

Cosmopolitan: An organism which is widely distributed, usually applied to species found on most continents, or most oceans if marine.

Cyanobacteria: Oxygenic photosynthetic bacteria, have often been referred to as blue green bacteria (or 'algae') in the past (see Margulis and Schwartz (1998)).

Cybernetics: The science of control systems, often used as a synonym for systems theory.

Detritivore: An organism which gains its energy from dead organic matter (detritus).

Diagenesis: Modifications to sediments, etc. which occur as they are changed into rock.

Diatoms: Aquatic single-cell eukaryotes with silica rich shells (see Margulis and Schwartz (1998)).

Diffraction grating: A collection of many microscopic grooves which reflects light in such a way that it is broken up into its constituent colours (see Parker (2005)).

Dinoflagellata: Group of planktonic protists ('protozoa'), also referred to as Dinomastigota (see Margulis and Schwartz (1998)).

Ecological Engineering: It has long been known that organisms modify their environment; however, during the 1980s Clive Jones and colleagues formalized this into the concept of 'ecological engineers'. These organisms have a particular effect on their environment, either by their physical structure (e.g. trees) or by their behaviour (e.g. beavers). Extinction of such species may have major effects on the environment.

Ecosystem services: Services of use to humans and other organisms provided by ecosystems, for example, oxygen, food, 'clean' water, etc.

Emergence: The formation of global patterns from local interactions; density dependent regulation is the most well-known ecological example (see Stewart (1998)). This concept was first named by the biologist G.H. Lewes (now better known as the partner of the novelist George Eliot) in the nineteenth century, but attracted relatively little interest until the late twentieth century (Ziman, 2003).

Encyst: Entering into a protective encapsulated form (often a dormant or 'resting' stage) usually in response to extreme environmental conditions.

Entropy: A measure of disorder of a system or substance, it can also be considered to be energy in a useless form (see Box 2.1 or Pielou (2001)).

Eukaryotes: One of the two great divisions ('Empires') of life on Earth, the other is the Prokaryotes (*q.v.*). They have a more complex cell structure than Prokaryotes usually with organelles such as mitochondria (see Margulis and Schwartz (1998)).

Food chain: A sequence describing feeding relationships. For example, Plant $\rightarrow$ Caterpillar $\rightarrow$ Tit (Chickadee) $\rightarrow$ Hawk, where the arrow shows feeding relationships. So in this example caterpillars eat plants but tits eat caterpillars (see Colinvaux (1980), chapter 3).

Food web: An interconnected network of foodchains (*q.v.*) describing the feeding relationships for a whole ecological community, such as a woodland or lake.

Free energy: Energy capable of doing useful work (see Pielou (2001)).

Gaia: The idea, suggested by James Lovelock, that organisms and their material environment evolve as a single-coupled system, from which emerges the sustained self-regulation of climate and chemistry at a habitable state for whatever is the current biota (see Section 1.4 of this book).

Geophysiology: A term sometimes used as a synonym for Gaia (*q.v.*).

Gray: A unit of absorbed radiation dose (1 Gray = 100 rad where 1 rad = 100 erg/g).

Group selection: Natural selection between groups (rather than individuals), in most cases this is a much less powerful force than individual-level selection. There is considerable semantic confusion in the literature over definitions of Group and Kin (*q.v.*) selection.

Guild: A group of species who all make their living in a similar way. Named for a perceived analogy to medieval guilds of tradesmen; however, the guilds of medieval England were rather more complex institutions than this analogy suggests.

Heat death: A state of uniform temperature (e.g. of a planet or the whole Universe) where entropy (*q.v.*) is maximal.

Holarctic: Biogeographical region, composed of Palaearctic (mid- and high-latitude Eurasia, including North Africa) and Nearctic (America, north of tropical Mexico).

Holocene: The past 10,000 years since the end of the last glaciation.

Homeostasis: The maintenance of a constant internal environment by an organism (such as human body temperature).

Host: The organism upon which a parasite (*q.v.*) lives, and usually, feeds.

Hyphae: The microscopic filaments which form the basic structure of a fungus (Spooner and Roberts, 2005).

Information: In the technical sense of Shannon information can be considered the reciprocal of entropy (*q.v.*) (see Box 2.1).

Kin selection: A special case of group selection (*q.v.*) where individuals are closely related—and so will have many genes in common.

Littoral: Living at the margin of a water body, for example, at the edge of a lake or in the intertidal zone of a sea.

Logistic equation: The simplest equation which relates the rate of growth of a population to that population's density so that $dN/dT = rN(1 - N/K)$, where N is population size, K is carrying capacity (*q.v.*), and r is the population growth rate.

Lotka–Volterra models: These model the competition between species based on the logistic equation (*q.v.*). So that the rate of population growth of species 1 when in competition with species 2 can be written as $dN_1/dT = r_1N_1 ((K_1 - N_1 - \alpha N_2)/K_1)$, where α is a competition coefficient and is effectively a way of expressing species 2 in units of species 1 (e.g. species 2 may be larger and require more resources than species 1), N, T, and K are as in the logistic equation.

Macroparasites: Multicellular parasites, such as nematodes and lice. Reproduce much more slowly than most microparasites (*q.v.*).

Magnetosphere: The volume of space around an astronomical body, such as a planet, where its magnetic field plays a prominent role.

Main sequence: Describes a group of stars on the Hurtzsprung–Russell diagram (a plot of stars surface temperature against absolute luminosity). Our sun is a main sequence star (Narlikar (1999), or any astronomy text book).

Mass ratio hypothesis: The idea that the extent to which a plant species affects ecosystem function shows a strong positive correlation with its biomass. This idea has been implicit in many discussions of plant ecology but was formalized by Philip Grime in the late 1990s. (see Section 6.5).

Mesocosm: A controlled experimental system, ideally closed to external influences. A small one would be a microcosm while a very large one would be a macrocosm, however, there are no generally accepted boundaries for when a microcosm becomes a mesocosm, etc.

Metamorphic: Rocks which have been altered by extreme heat and/or pressure.

Metazoa: Animals excluding 'protozoa' like organisms.

Microparasites: Small parasitic organisms such as viruses, bacteria, and protozoa. Often capable of extreme multiplication within the host (*q.v.*) as they have very short generation times. *c f.* macroparasites.

Mutualism: A mutually beneficial relationship, usually used to describe the relationship between different species. See also symbiosis (*q.v.*).

Mycorrhizae: An association between plant roots and fungi which is normally beneficial to both species (Wakeford (2001), chapter 2).

Niche construction: The process by which organisms modify their own, or other organisms environments. This approach was developed during the 1990s by John Odling-Smee, Kevin Laland, and Marcus Feldman, it emphasizes the evolutionary effects of this process and has its theoretical origins in ideas from population genetics (for a nontechnical account see Jones (2005)).

Occult precipitation: Deposition of water directly from clouds, for example, by deposition onto tree branches etc.—also called 'cloudwater deposition' or 'horizontal deposition'.

Organic chemistry: The chemistry of carbon compounds, note this does not imply that they are of biological origin. For example, many organic chemicals have now been identified in space, however, the majority of astronomers believe all of these are of abiological origin.

Palaeosol: An ancient soil, may have undergone lithification (i.e. been turned to rock).

Palaeozoic: Geological era covering the period of time between 542 and 251 million years ago.

Panspermia: Literally 'seeds everywhere', the idea that life arrives on a planet from elsewhere in the Universe. A variant is the concept of directed panspermia (an idea particularly associated with Francis Crick and Leslie Orgel) where microbial life is deliberately spread through space by an intelligent species (Crick, 1982; Wickramasinghe, 2005).

Parasite: An organism which lives in, or on, another organism and in so doing obtains resources at its host's (*q.v.*) expense. It is remarkably difficult to produce a rigorous definition of parasite which successfully covers the wide range of examples seen in nature (Box 3.2). See also microparasite and macroparasite.

Parasitoid: An organism which lays its eggs inside another organism, the growth of the parasitoid ultimately kills the host. The classic examples are species of wasps and flies.

Partial pressure: In a mixture of different gasses the partial pressure is the pressure that a gas would exert if it occupied that volume by itself.

Pelagic: Living in open water.

Phagocytosis: The engulfing of solid particles by a cell by the budding off of a vesicle from the cell membrane.

Phanerozoic: The geological eon covering the most recent 542 million years of Earth history (from the start of the Cambrian until the present). For its subdivision into geological periods see the table at the end of this glossary.

Photorespiration: The competition between O_2 and CO_2 for sites on the enzyme rubisco, this process can reduce the efficiency of photosynthesis.

Photosystem I and II: Two light-capturing molecular systems involved in the 'light reaction' of photosynthesis.

Predator: An organism which consumes another killing it in the process. It is remarkably difficult to produce a rigorous definition of predator which successfully distinguishes it from parasite (*q.v.*) (see Box 3.2).

Primary production: The rate of production of biomass (*q.v.*) by autotrophs (*q.v.*) (see Pielou (2001), chapter 11).

Primary succession: The process by which life colonizes bare ground, (*q.v.*) 'Succession'.

Prokaryotes: Single-celled organisms lacking a nucleus and other organelles, bacteria *sensu lato*. One of the two great divisions ('Empires') of life on Earth, the other is the Eukaryotes (*q.v.*).

Proterozoic: Geological eon from 542 to 2,500 million years ago (the late 'Precambrian').

Pseudopodia: Temporary projections of cellular material in some protozoan cells, used in feeding or locomotion.

Quadrat: An area, often square, marked out on the ground to help record the distribution or abundance of species.

Reactive oxygen species: Oxygen containing molecules with an unpaired electron (free radicals).

Reduction: In a chemical context a reaction in which a molecule has oxygen removed, or hydrogen or electrons added; also the addition of electrons to an atom. In philosophy the act of explaining concepts or processes at one level in the terms of those from another level, usually seen as more basic. The often heard statement that 'all biology is just chemistry' is an example of a reductive approach to biology—and ignores crucial ideas such as natural selection or adaptation which are not present in chemistry.

Second Law of Thermodynamics: At its simplest this states that heat flows from a hot body to a cold one, more formally than the free energy (*q.v.*) of a system decreases in any spontaneous change (see Section 2.1 of this book or Pielou (2001)).

Snowball Earth: The idea, particularly associated with Paul Hoffman, that at various times in the geological past the Earth has suffered major ice ages which lead to most of the planet freezing. People who argue for less extreme events often write about a 'slush ball Earth' (see Kump *et al.* (2004), chapter 12).

Stefan Boltzmann constant: These two physicists showed that the energy given off by a perfect or 'black body' radiator is: $B = \sigma T^4$, where T is temperature and σ is the Stefan Boltzmann constant which is 5.67×10^{-8} Wm^{-2}K^{-1}. In the context of the Daisyworld model it is relevant that at within the range of temperatures typical of the Earth's surface, soil, and vegetation approximate to black body radiators (Monteith and Unsworth, 1990).

Stomata: Opening in the surface of a plant (often in the leaves) which allows gaseous exchange between the plant's interior and the surrounding atmosphere.

Stromatolites: Laminated (sometimes domed) structures in rocks caused by the trapping and binding of sediment particles by microorganisms.

Succession: A directional change in ecological communities over time, most often used in the context of vegetation change (Moore (2001), for an excellent short account also Colinvaux (1980), chapter 12).

Supernova: An 'exploding' star caused by the contraction of its core (see Narlikar (1999), chapter 3).

Symbiosis: In its original usage this described two organisms living very closely together. Such a relationship could be mutually beneficial, harmful to one of the partners or neutral with respect to benefits. This is the way the term is used in this book, however, some authors use it as a synonym for mutualism (*q.v.*) (see Wilkinson (2001b), for a discussion of these alternative definitions). Wakeford (2001) is a highly readable 'popular' book on the subject.

Teleological: Explanations that assume a purpose, for example, explanations of organisms adaptations as designed for the purpose by a God. Such explanations are rightly considered highly suspect by science, as such suggestions that ideas like Gaia (*q.v.*) or Maximum Entropy Production (Section 2.4) imply such a purpose has made many scientists treat them with suspicion.

Testate amoebae: Shell-building protozoa which are often common in soils and freshwater, especially in sediments of high organic matter content (see Fig. 4.1).

Thermodynamics: The study of energy transformations (see Pielou (2001), for a simple introduction to the thermodynamics of the Earth system).

Xylem: Vascular woody plant tissue which allows the transport of water—along with dissolved minerals—around a plant.

The geological time scale

The geological time scale for the Earth is split into four 'eons' which are subdivided into 'eras', these eras are further subdivided into the more familiar geological periods (systems). In assigning dates to this time scale I have followed the recent (2004) recommendations of the International Commission on Stratigraphy (summarized by Whitfield (2004)). The one exception to this is that I have retained the Quaternary as a full geological period starting 2.6 million years ago, the formal status of the Quaternary in the geological time scale is currently controversial (see Pillans and Naish (2004) for a discussion and Giles (2005) for a more journalistic account), however, its distinctive climate which has important ecological implications argues for its retention as a useful subdivision of geological time.

Eons—see also within the glossary above.

Hadean: Before approximately 3,800 million years ago

Archaean: From approximately 3,800 to 2,500 million years ago.

Proterozoic: From 2,500 to 542 million years ago

Phanerozoic: From 542 million years ago until the present.

The Phanerozoic in more detail

Within this book I only use formal names of eras and periods (systems) within the most recent geological era the Phanerozoic. Note that the well-known period of the 'Tertiary' no longer has formal status and is split into the Palaeogene and the Neogene—this does not stop most geologists talking about the 'Cretaceous/Tertiary boundary' when discussing mass extinctions etc. as old habits die hard. The subdivisions of the Phanerozoic are shown in the following table:

Era	Period (system)	Age in years (millions of years)
Cenozoic	Quaternary	0–2.6*
Cenozoic	Neogene	2.6–23.03
Cenozoic	Palaeogene	23.03–65.5
Mesozoic	Cretaceous	65.5–145.5
Mesozoic	Jurassic	145.5–199.6
Mesozoic	Triassic	199.6–251.0
Palaeozoic	Permian	251.0–299.0
Palaeozoic	Carboniferous	299.0–359.2
Palaeozoic	Devonian	359.2–416.0
Palaeozoic	Silurian	416.0–443.7
Palaeozoic	Ordovician	443.7–488.3
Palaeozoic	Cambrian	488.3–542.0

*Note many authors use 1.8 million as the base of the Quaternary, however 2.6 seems to be gaining increasing support (Pillans and Naish, 2004).

References

Abraham, E.R., Law, C.S., Boyd, P.W., Lavender, S.J., Maldonado, M.T. and Bowie, A.R. 2000. Importance of stirring in the development of an iron-fertilized phytoplankton bloom. *Nature*, **407**, 727–30.

Ackland, G.J., Clark, M.A. and Lenton, T.M. 2003. Catastrophic desert formation in Daisyworld. *Journal of Theoretical Biology*, **223**, 39–44.

Adams, B., Carr, J., Lenton, T.M. and White, A. 2003. One-dimensional daisyworld: spatial interactions and pattern formation. *Journal of Theoretical Biology*, **233**, 505–13.

Aleksander, I. 2002. Understanding information, bit by bit. In *It must be beautiful great equations of modern science* (ed. Farmelo, G), pp. 131–48. London: Granta Books.

Allee, W.C., Emerson, A.E., Park, O. and Schmidt, K.P. 1949. *Principles of Animal Ecology*. Philadelphia: W.B. Saunders.

Allen, A.P., Gillooly, J.F. and Brown, J.H. 2005. Linking the global carbon cycle to individual metabolism. *Functional Ecology*, **19**, 202–13.

Allison, S.D. 2005. Cheaters, diffusion and nutrients constrain decomposition by microbial enzymes in spatially structured environments. *Ecology Letters* **8**, 626–35.

Alvarez, W. 1997. *T. rex and the crater of doom*. Princeton: Princeton University Press.

Anderson, R.M. and May, R.M. 1991. *Infectious diseases of humans, dynamics and control*. Oxford: Oxford University Press.

Archibold, O.W. 1995. *Ecology of world vegetation*. London: Chapman and Hall.

Arrigo, K.R. 2005. Marine microorganisms and global nutrient cycling. *Nature*, **437**, 349–55.

Arroyo-Mora, J.P., Sánchez-Azofeifa, G.A., Rivard, B., Calvo, J.C. and Janzen, D.H. 2005. Dynamics in landscape structure and composition for the Chorotega region, Costa Rica from 1960 to 2000. *Agriculture, Ecosystems and Environment*, **106**, 27–39.

Ashmole, P. and Ashmole M. 2000. *St Helena and Ascension Island: a natural history* Oswestry: Anthony Nelson.

Atlas, R.M. and Bartha, R. 1998. *Microbial ecology, fundamentals and applications* 4th ed. Menlo Park, California: Benjamin Cummins.

Ball, P. 2005a. Seeking the solution. *Nature*, **436**, 1084–5.

Ball, P. 2005b. *Elegant solutions*. Cambridge: Royal Society of Chemistry.

Barnes, D.K.A. 2002. Invasions by marine life on plastic debris. *Nature*, **416**, 808–9.

Beerling, D.J. 2005. Leaf evolution: gases, genes and geochemistry. *Annals of Botany*, **96**, 345–52.

Beerling, D.J. and Berner, R.A. 2005. Feedbacks and the coevolution of plants and atmospheric CO_2. *Proceedings of the National Academy of Sciences USA*, **102**, 1302–5.

Beerling, D.J., Chaloner, W.G., Huntley, B., Pearson, J.A. and Tooley, M.J. 1993. Stomatal density responds to the glacial cycle of environmental change. *Proceedings of the Royal Society London B.*, **251**, 133–8.

Beerling, D.J., Lake, J.A., Berner, R.A., Hickey, L.J., Taylor, D.W. and Royer, D.L. 2002a. Carbon isotope evidence implying high O_2/CO_2 ratios in the Permo-Carboniferous atmosphere. *Geochimica et Cosmochimica Acta*, **66**, 3757–67.

Beerling, D.J., Lomax, B.H., Upchurch Jr., G.R., Nichols, D.J., Pillmore, C.L., Handley, L.L. and Scrimgeour, C.M. 2001a. Evidence for the recovery of terrestrial ecosystems ahead of marine primary production following a biotic crisis at the Cretaceous-Tertiary boundary. *Journal of the Geological Society, London*, **158**, 737–40.

Beerling, D.J., Lomax, B.H., Royer, D.L., Upchurch Jr., G.R. and Kump, L.R. 2002b. An atmospheric pCO_2 reconstruction across the Cretaceous-Tertiary boundary from leaf megafossils. *Proceedings of the National Academy of Sciences USA*, **99**, 7836–40.

Beerling, D.J., Osborne, C.P. and Chaloner, W.G. 2001b. Evolution of leaf-form in land plants linked to atmospheric CO_2 decline in the Late Palaeozoic era. *Nature*, **410**, 352–4.

Begon, M., Townsend, C.R. and Harper, J.L. 2006. *Ecology: individuals, populations and communities*. 4th ed. Oxford: Blackwells.

Belcher, C.M., Collinson, M.E. and Scott, A.C. 2005. Constraints on the thermal energy released from the Chicxulub impactor: evidence from multi-method charcoal analysis. *Journal of the Geological Society London*, **162**, 591–602.

Bennett, K.D. 1989. A provisional map of forest types for the British Isles 5000 years ago. *Journal of Quaternary Science*, **4**, 141–4.

Bennett, K.D. 1997. *Evolution and ecology*. Cambridge: Cambridge University Press.

Bergman, N.M., Lenton, T.M. and Watson, A.J. 2004. COPSE: a new model of biogeochemical cycling over Phanerozoic time. *American Journal of Science*, **304**, 397–437.

Berkner, L.V. and Marshall, L.C. 1972. Oxygen and evolution. In *Understanding the Earth* 2nd ed. (ed. Glass, I.G., Smith, P.J. and Wilson, R.C.L.) pp. 143–9. Sussex: The Artimis Press.

Berner, R.A. 1998. The carbon cycle and CO_2 over Phanerozoic time: the role of land plants. *Philosophical Transactions of the Royal Society London B.*, **353**, 75–82.

Berner, R.A. 1999. Atmospheric oxygen over Phanerozoic time. *Proceedings of the National Academy of Sciences USA*, **96**, 10955–7.

Berner, R.A. 2004. *The Phanerozoic carbon cycle*. Oxford: Oxford University Press.

Bessa-Gomes, C., Legendre, S. and Clobert, J. 2004. Allee effect, mating systems and the extinction risk in populations with two sexes. *Ecology Letters*, **7**, 802–12.

Betts, R.A. 1999. Self-beneficial effects of vegetation on climate in an Ocean-Atmosphere General Circulation Model. *Geophysical Research*, **26**, 1457–60.

Betts, R.A. 2000. Offset of the potential carbon sink from boreal forestation by decreases in surface albedo. *Nature*, **408**, 187–90.

Bond, W.J. and Keeley, J.E. 2005. Fire as a global 'herbivore': the ecology and evolution of flammable ecosystems. *Trends in Ecology and Evolution*, **20**, 387–94.

Bourne, W.R.P. and Simmons, K.E.L. 1998. A preliminary list of the birds of Ascension Island, South Atlantic Ocean. *Sea Swallow*, **47**, 42–56.

Bradshaw, A.D. 1987. Comparison—its scope and limits. *New Phytologist*, **106**, 3–21.

Bradshaw, R.H.W. 1999. Spatial response of animals to climate change during the Quaternary. *Ecological Bulletins* **47**, 16–21.

Brasier, M.D. 1979. The early fossil record. *Chemistry in Britain*, **15**, 588–92.

Brasier, M.D., Green, O.R., Jephcoat, A.P., Kleppe, A.K., Van Kranendonk, M.J., Lindsay, J.F., Steele, A. and Grassineau, N.V. 2002. Questioning the evidence for Earth's oldest fossils. *Nature*, **416**, 76–81.

Brocks, J.J., Love, G.D., Summons, R.E., Knoll, A.H., Logan, G.A. and Bowden, S.A. 2005. Biomarker evidence for green and purple sulphur bacteria in a stratified Palaeoprotozoic sea. *Nature*, **437**, 866–70.

Brook, B.W. and Kikkawa, J. 1998. Examining threats faced by island birds: a population viability analysis on the Capricorn Silvereye using long-term data. *Journal of Applied Ecology*, **35**, 491–503.

Buckland, P.C. and Edwards, K.J. 1984. The longevity of pastoral episodes of clearance activity in pollen diagrams: the role of post-occupation grazing. *Journal of Biogeography*, **11**, 243–9.

Buckling, A., Kassen, R., Bell, G. and Rainey, P.B. 2000. Disturbance and diversity in experimental microcosms. *Nature*, **408**, 961–4.

Buckling, A. and Rainey, P.B. 2002. The role of parasites in sympatric and allopatric host diversification. *Nature*, **420**, 496–9.

Bush, M.B. and Colinvaux, P.A. 1994. Tropical forest disturbance: paleoecological records from Darien, Panama. *Ecology*, **75**, 1761–8.

Cain, M.L., Damman, H. and Muir, A. 1998. Seed dispersal and the Holocene migration of woodland herbs. *Ecological Monographs*, **68**, 325–47.

Calow, P (ed) 1999. *Concise Encyclopaedia of Ecology*. Oxford: Blackwell.

Canfield, D.E. 1999. A breath of fresh air. *Nature*, **400**, 503–5.

Cano, R.J. and Borucki, M.K. 1995. Revival and identification of bacterial spores in 25- to 40-Million-year-old Dominican Amber. *Science*, **268**, 1060–4.

Capone, D.G., Burns, J.A., Monotoya, J.P., Subramaniam, A., Mahaffey, C., Gunderson, T., Michaels, A.F. and Carpenter, E.J. 2005. Nitrogen fixation by *Trichodesmium spp*: An important source of new nitrogen to the tropical and subtropical Atlantic Ocean. *Global Biogeochemical Cycles*, **19**, GB2024, doi: 10.1029/2004GB002331.

Carr, B.J. and Rees, M.J. 1979. The anthropic principle and the structure of the physical world. *Nature*, **278**, 605–12.

Carroll, S.B. 2001. Chance and necessity: the evolution of morphological complexity and diversity. *Nature*, **409**, 1102–9.

Carter, R.N. and Prince, S.D. 1981. Epidemic models used to explain biogeographical distribution limits. *Nature*, **293**, 644–5.

Case, T.J. 2000. *An illustrated guide to theoretical ecology*. Oxford: Oxford University Press.

Catling, D.C. and Claire, M.W. 2005. How Earth's atmosphere evolved to an oxic state: a status report. *Earth and Planetary Science Letters*, **237**, 1–20.

Caughley, G. 1994. Directions in conservation biology. *Journal of Animal Ecology*, **63**, 215–44.

Cavalier-Smith, T. 2004. Only six kingdoms of life. *Proceedings of the Royal Society of London B*, **271**, 1251–62.

Chambers, F.M. 1995. Climate response, migrational lag, and pollen representation: the problems posed by *Rhododendron* and *Acer. Historical Biology*, **9**, 243–56.

Chambers, F.M. and Brain, S.A. 2002. Paradigm shifts in late-Holocene climatology? *The Holocene*, **12**, 239–49.

Charlson, R.J., Lovelock, J.E., Andreae, M.O. and Warren, S.G. 1987. Oceanic phytoplankton, atmospheric sulphur, cloud albedo and climate. *Nature*, **326**, 655–61.

Cherrett, J.M. 1989. Key concepts: the results of a survey of our members' opinions. In *Ecological concepts*. (ed. Cherrett, J.M.), pp. 1–16. Oxford: Blackwell.

Chyba, C.F. 2000. Energy for microbial life on Europa. *Nature*, **403**, 381–2.

Chyba, C.F. and Phillips, C.B. 2001. Possible ecosystems and the search for life on Europa. *Proceedings of the National Academy of Sciences USA*, **98**, 801–4.

Clare, T. 1996. Archaeology, conservation and the late twentieth-century village landscape, with particular reference to Westmorland and North Lancashire in northwestern England. *Conservation and Management of Archaeological Sites*, **1**, 169–88.

Clymo, R.S., Turunen, J. and Tolonen, K. 1998. Carbon accumulation in peatland. *Oikos*, **81**, 368–88.

Cockell, C.S. and Blaustein, A.R. 2000. "Ultraviolet spring" and the ecological consequences of catastrophic impacts. *Ecology Letters*, **3**, 77–81.

Coe, M. 1987. Unforeseen effects of control. *Nature*, **327**, 367.

Cohen, J.E. 1995. *How many people can the Earth support?* New York: W.W. Norton.

Cohen, J.E. 2004. Mathematics is Biology's next microscope, only better; Biology is Mathematics next Physics only better. *PLoS Biology*, **2**, e439.

Cohen, J.E., Jonsson, T. and Carpenter, S.R. 2003. Ecological community description using the food web, species abundance, and body size. *Proceeding of the National Academy of Sciences USA*, **100**, 1781–6.

Cohen, J.E., Pimm, S.L., Yodzis, P. and Saldana, J. 1993. Body sizes of animal predators and animal prey in food webs. *Journal of Animal Ecology*, **62**, 67–78.

Cohen, J.E. and Rich, A.D. 2000. Interspecific competition affects temperature stability in Daisyworld. *Tellus*, **52B**, 980–4.

Cohen, J.E. and Tilman, D. 1996. Biosphere 2 and biodiversity: the lessons so far. *Science*, **274**, 1150–1.

Coles, B. 2001. The impact of beaver activity on stream channels: some implications for past landscapes and human activity. *Journal of Wetland Archaeology*, **1**, 55–82.

Colinvaux, P. 1980. *Why big fierce animals are rare*. London: Penguin.

Colinvaux, P. 1993. *Ecology 2*. New York: Wiley.

Connell, J.H. 1961. The influence of interspecific competition and other factors on the distribution of the barnacle *Chthamalus stellatus*. *Ecology*, **42**, 710–23.

Conway Morris S. 1989. Burgess Shale faunas and the Cambrian explosion. *Science*, **246**, 339–45.

Conway Morris, S. 1998, *The crucible of creation*. Oxford: Oxford University Press.

Conway Morris S. 2000. The Cambrian "explosion": slow-fuse or megatonnage? *Proceedings of the National Academy of Science USA*, **97**, 4426–9.

Conway Morris, S., Peel, J.S., Higgins, A.K., Soper, N.J. and Davis, N.C. 1987. A Burgess shale-like fauna from the Lower Cambrian of North Greenland. *Nature*, **326**, 181–3.

Coombs, F. 1978. *The crows, a study of the corvids of Europe*. London: B.T. Batsford.

Cotgreave, P. and Forseth, I. 2002. *Introductory Ecology*. Oxford: Blackwell.

Cott, H.B. 1940. *Adaptive coloration in animals*. London: Methuen.

Coulson, T., Mace, G.M., Hudson, E. and Possingham, H. 2001. The use and abuse of population viability analysis. *Trends in Ecology and Evolution*, **16**, 219–21.

Courchamp, F., Clutton-Brock, T. and Grenfell, B. 1999. Inverse density dependence and the Allee effect. *Trends in Ecology and Evolution*, **14**, 405–10.

Cox, P.M., Betts, R.A., Jones, C.D., Spall, S.A. and Totterdale, L.J. 2000. Acceleration of global warming due to carbon-cycle feedbacks in a coupled climate model. *Nature*, **408**, 184–7.

Cramp, S. and Perrins, C.M. (eds) 1994. *The birds of the Western Palearctic. Vol 8*. Oxford: Oxford University Press.

Cramp, S. and Simmons, K.E.L. (eds) 1979. *The birds of the Western Palaearctic Vol. 2*. Oxford: Oxford University Press.

Crawford, R.M.M. 1989. *Studies in plant survival*, Oxford: Blackwell Science.

Crick, F. 1982. *Life itself*. London: Macdonald & Co.

Crist, E. 2004. Concerned with trifles? A geophysiological reading of Charles Darwin's last book. In *Scientists debate Gaia*. (ed. Schneider, S.H., Miller, J.R., Crist, E. and Boston, P.J.) pp. 161–72. Cambridge Massachusetts: MIT Press

Cross, J.R. 1975. Rhododendron ponticum (Biological Flora of the British Isles). *Journal of Ecology*, **63**, 345–64.

Crutzen, P.J. and Steffen, W. 2003. How long have we been in the anthropocene era? *Climatic Change*, **61**, 251–7.

Culley, A.I., Lang, A.S. and Suttle, C.A. 2003. High diversity of unknown picorna-like viruses in the sea. *Nature*, **424**, 1054–7.

Cushman, J.C. 2004. Photosynthesis to crassulacean acid metabolism shift in Mesembryanthenum crystallinum: a stress tolerance mechanism. In *Encyclopaedia of Plant and Crop Science*. (ed. Goodman, R.M.) pp. 241–4. New York: Marcel Dekker.

Dang, C.K., Chauvert, E. and Gessner, M.O. 2005. Magnitude and variability of process rates in fungal diversity-litter decomposition relationships. *Ecology Letters*, **8**, 1129–37.

Darwin, C. 1859. *On the origin of species by means of natural selection*. London: John Murray.

Darwin, C. 1875. *Insectivorous Plants*. London: John Murray.

Darwin, C. 1881. *The formation of vegetable mould through the action of worms, with observations on their habits*. London: John Murray.

Davies, K.M. 2004. Phenylpropanoids. In *Encyclopaedia of Plant and Crop Science*. (ed. Goodman, R.M.) pp. 868–71. New York: Marcel Dekker.

Davies, P. 1995. *Are we alone?* London: Penguin.

Davies, P. 1998. *The fifth miracle*. London: Allen Lane.

Davies, P. 2005. A quantum recipe for life. *Nature*, **437**, 819.

Davis, S.R. and Wilkinson, D.M. 2004. The conservation management value of testate amoebae as 'restoration' indicators: speculations based on two damaged raised mires in northwest England. *The Holocene*, **14**, 135–43.

Dawkins, R. 1982. *The extended phenotype*. Oxford: Oxford University Press.

de Castro, F. and Bolker, B. 2005. Mechanisms of disease-induced extinction. *Ecology Letters* **8**, 117–26.

Dehnen-Schmutz, K., Perring, C. and Williamson, M. 2004. Controlling *Rhododendron ponticum* in the British Isles: an economic analysis. *Journal of Environmental Management*, **70**, 323–32.

de la Bédoyère, G. 2002. *Gods with thunderbolts, religion in Roman Britain*. Stroud: Tempus.

del Hoyo, J., Elliot, A. and Sargatal, J. (eds) 1992. *Handbook of the birds of the world. Vol 1*. Barcelona: Lynx Editions.

del Hoyo, J., Elliot, A. and Sargatal, J. (eds) 1994. *Handbook of the birds of the world. Vol 2*. Barcelona: Lynx Editions.

Des Marais D.J. 2005. Sea change in sediments. *Nature*, **437**, 826–7.

Diamond, J. 1997. *Guns, germs and steel*. London: Jonathan Cape.

Diamond, J. 2002. Evolution, consequences and future of plant and animal domestication. *Nature*, **418**, 700–7.

Doak, D.F., Bigger, D., Harding, E.K., Marvier, M.A., O'Mally, R.E. and Thomson, D. 1998. The statistical inevitability of stability-diversity relationships in community ecology. *American Naturalist*, **151**, 264–76.

Dobbs, B.J.T. 1982. Newton's alchemy and his theory of matter. *Isis*, **73**, 511–28.

Dobson, C.M. 2004. Chemical space and biology. *Nature*, **432**, 824–8.

Downing, K. 2000. Exploring Gaia theory: artificial life on a planetary scale. In *Proceedings of the 7th International Conference on Artificial Life*. (ed. Bedau, M., McCaskill, J., Parkard, N. and Rasmussen, S.) pp. 90–9. Cambridge Mass: MIT Press.

Downing, K.L. 2002. The simulated emergence of distributed environmental control in evolving microcosms. *Artificial Life*, **8**, 123–53.

Downing, K.L. 2004. Gaia in the machine: the artificial life approach. In *Scientists debate Gaia*. (ed. Schneider, S.H., Miller, J.R., Crist, E. and Boston, P.J.) pp. 267–80.Cambridge Massachusetts: MIT Press.

Downing, K and Zvirinsky, P. 1999. The simulated evolution of biochemical guilds: reconciling Gaia theory and natural selection. *Artificial Life*, **5**, 291–318.

Du Rietz, G.E. 1957. Linne som myrforskare. *Uppsala Universitets Arsskrift*, **5**, 1–80.

Dyer, J.M. 1995. Assessment of climatic warming using a model of forest species migration. *Ecological Modelling*, **79**, 199–219.

Edlin, H.L. 1966. *Trees, woods and man*. 2nd ed. London: Collins.

Ehrlich, P. 1991. Coevolution and its applicability to the Gaia hypothesis. In *Scientists on Gaia*. (ed. Schneider, S.H. and Boston, P.J.) pp. 19–22. Cambridge Massachusetts: MIT Press.

Eigen, M. and Schuster, P. 1977. The hypercycle. A principle of natural self-organisation part A: emergence of the hypercycle. *Naturwissenschaften*, **58**, 465–523.

Elton, C. 1927. *Animal Ecology*. London: Sidgwick and Jackson.

Elton, C.S. 1958. *The ecology of invasion by animals and plants*. London: Chapman and Hall.

Erwin, T.L. 1982. Tropical forests: their richness in Coleoptera and other arthropod species. *Coleopterists Bulletin*, **36**, 74–5.

Falkowski, P.G. 2002. The oceans invisible forest. *Scientific American*, **287**, 38–45.

Falkowski, P.G. and Davis, C.S. 2004. Natural proportions. *Nature*, **431**, 131.

Fenchel, T. 1992. What can ecologists learn from microbes: life beneath a square centimetre of sediment surface. *Functional Ecology*, **6**, 499–507.

Fenchel, T. 2005. Cosmopolitan microbes and their 'cryptic' species. *Aquatic Microbial Ecology*, **41**, 49–54.

Fenchel, T. and Finlay, B.J. 1995. *Ecology and evolution in anoxic worlds*. Oxford: Oxford University Press.

Fenchel, T., King, G.M. and Blackburn, T.H. 1998. *Bacterial biogeochemistry* 2nd ed. San Diego: Academic Press.

Finlay, B.J. 2002. Global dispersal of free-living microbial eukaryote species. *Science*, **296**, 1061–3.

Finlay, B.J. and Clarke, K.J. 1999. Ubiquitous dispersal of microbial species. *Nature*, **400**, 828.

Finlay, B.J., Esteban, G.F. and Fenchel, T. 2004. Protist diversity is different? *Protist*, **155**, 15–22.

Finlay, B.J., Marberly, S.C. and Cooper, J.I. 1997. Microbial diversity and ecosystem function. *Oikos*, 80, 209–13.

Fisher, R.A. 1930. *The genetical theory of natural selection*. Oxford: The Clarendon Press.

Floate, K.D., Wardhaugh, K.G., Boxall, A.B.A. and Sherratt, T.N. 2005. Fecal residues of veterinary parasiticides: nontarget effects in the pasture environment. *Annual Review of Entomology*, **50**, 153–79.

Foissner, W. 1998. An updated compilation of world soil ciliates (Protozoa, Ciliophora), with ecological notes, new records, and descriptions of new species. *European Journal of Protistology*, **34**, 195–235.

Foissner, W. 1999. Protist diversity: estimates of the near-imponderable. *Protist*, **150**, 363–8.

Förster, B. and Pogson, B.J. 2004. Carotenoids in photosynthesis. In *Encyclopaedia of Plant and Crop Science*. (ed. Goodman, R.M.) pp. 245–9. New York: Marcel Dekker.

Foster, D.R. 1992. Land-use history (1730–1990) and vegetation dynamics in central New England, USA. *Journal of Ecology*, **80**, 753–72.

Franck, S., Von Bloh, W., Bounana, C. and Schellnhuber, H-J. 2004. Extraterrestrial Gaias. In *Scientists debate Gaia*. (ed. Schneider, S.H., Miller, J.R., Crist, E. and Boston, P.J.) pp. 309–19.Cambridge Massachusetts: MIT Press.

Frank, S.A. 1998. *Foundations of social evolution*. Princeton: Princeton University Press.

Fuhrman, J. 2003. Genome sequences from the sea. *Nature*, **424**, 1001–2.

Fuller, E. 2000. *Extinct birds*. Oxford: Oxford University Press.

Gattuso, J.-P. and Buddemeier, R.W. 2000. Calcification and CO_2. *Nature*, **407**, 311–13.

Gilbert, D., Amblard, C., Bourdier, G., Francez, A.-J. and Mitchell, E.A.D. 2000. Le regime alimentaire des Thecamoebiens (Protista, Sarcodina). *Annee Biologique*, **39**, 57–68.

Gilbert, D., Mitchell, E.A.D., Amblard, C., Bourdier, G. and Francez, A.-J. 2003. Population dynamics and food preferences of the testate amoebae *Nebela tincta major-bohemica-collaris* complex (Protozoa) in a Sphagnum peatland. *Acta Protozoologica*, **42**, 99–104.

Giles, J. 2005. Geologists call time on dating dispute. *Nature*, **435**, 865.

Gillon, J. 2000. Feedback on Gaia. *Nature*, **406**, 685–6.

Givnish, T.J. 1987. Comparative studies of leaf form: assessing the relative roles of selective pressures and phylogenetic constraints. *New Phytologist*, **106**, 131–60.

Glen, W. (ed.) 1994. *Mass-extinction debates: how science works in a crisis*. Stanford: Stanford University Press.

Godwin, H. 1975. *History of the British flora*. Cambridge: Cambridge University Press.

Gold, T. 1999. *The deep hot biosphere*. New York: Springer-Verlag.

Goodwin, D. 1986. *Crows of the world* 2nd ed. London: British museum (Natural History).

Gould, S.J. 1990. *Wonderful life*. London: Hutchinson Radius.

Gould, S.J. 1995. 'What is life?' as a problem in history. In *What is life? The next fifty years*. (ed. Murphy, M.P. and O'Neill, L.A.J.) pp. 25–39. Cambridge: Cambridge University Press.

Gould, S.J. 2002. *The structure of evolutionary theory*. Cambridge Massachusetts: Harvard University Press.

Grace, J. 1997. Plant water relations. In *Plant ecology* 2nd ed. (ed Crawley, M.J.) pp. 28–50. Oxford: Blackwell.

Grace, J. 2004. Understanding and managing the global carbon cycle. *Journal of Ecology*, **92**, 189–202.

Grady, M.M., Hutchison, R., McCall, G.J.H. and Rothery, D.A. 1998. *Meteorites: flux with time and impact factors*. London: Geological Society, Special Publication **140**.

Gregory, J. 2005. *Fred Hoyle's universe*. Oxford: Oxford University Press.

Gribaldo, S. and Philippe, H. 2002. Ancient phylogenetic relationships. *Theoretical Population Biology* **61**, 391–408.

Gribbin, J. 2000. *Stardust*. London: Allen Lane.

Grieve, R.A.F. 1998. Extraterrestrial impacts on Earth: the evidence and the consequences. In *Meteorites: Flux with time and impact effects*. (ed. Grady, M.M., Hutchinson, R., McCall, G.J.H. and Rothery, D.A.). Geological Society, London: Special Publications, **140**, 105–31.

Grime, J.P. 1974. Vegetation classification by reference to strategies. *Nature*, **250**, 26–31.

Grime, J.P. 1979. *Plant strategies and vegetation processes*. Chichester: John Wiley.

Grime, J.P. 1997. Biodiversity and ecosystem function: the debate deepens. *Science*, **277**, 1260–1.

Grime, J.P. 1998. Benefits of plant diversity to ecosystems: immediate, filter and founder effects. *Journal of Ecology*, **86**, 902–10.

Grime, J.P. 2001. *Plant strategies, vegetation processes and ecosystem properties.* Chichester: John Wiley.

Haila, Y. 1999a. Socioecologies. *Ecography*, **22**, 337–48.

Haila, Y. 1999b. Biodiversity and the divide between culture and nature. *Biodiversity and Conservation*, **8**, 165–81.

Haldane, J.B.S. 1932. *The causes of evolution.* London: Longmans, Green & Co.

Hale, W.G., Margham, J.P. and Saunders, V.A. 2005. *Collins Dictionary of Biology* 3rd ed. London: Collins.

Hall, D., Wells, C.E. and Huckerby, E. 1995. *The wetlands of Greater Manchester.* Lancaster: Lancaster Imprints.

Hall, S.J. and Raffaelli, D. 1991. Food-web patterns: lessons from a species rich web. *Journal of Animal Ecology*, **60**, 823–41.

Hamilton, W.D. 1971. Geometry for the selfish herd. *Journal of Theoretical Biology*, **31**, 295–311.

Hamilton, W.D. 1995. Ecology in the large: Gaia and Genghis Khan. *Journal of Applied Ecology*, **32**, 415–53.

Hamilton, W.D. 2001. *Narrow roads of gene land. Vol 2.* Oxford: Oxford University Press.

Harcourt, A.H. 1995. Population viability estimates: theory and practice for a wild Gorilla population. *Conservation Biology*, **9**, 134–42.

Harding, R.J. and Pomeroy, J.W. 1996. The energy balance of the winter Boreal landscape. *Journal of Climatology*, **9**, 2778–87.

Harding, S.P. 1999. Food web complexity enhances community stability and climate regulation in a geophysical model. *Tellus*, **51B**, 815–29.

Harding, S.P. and Lovelock, J.E. 1996. Exploiter-mediated coexistence and frequency-dependent selection in a numerical model of biodiversity. *Journal of Theoretical Biology*, **182**, 109–16.

Hardy, A. 1956. *The Open Sea: the world of Plankton.* London: Collins.

Hazen, R.M. and Roedder, E. 2001. How old are bacteria from the Permian age? *Nature*, **411**, 155.

Heimann, M. 2005. Charles David Keeling 1928–2005. *Nature*, **437**, 331.

Henry, J. 2002. *The scientific revolution and the origins of modern science.* 2nd ed. London: Palgrave.

Higgins, S.I., Nathan, R. and Cain, M.L. 2003. Are long distance dispersal events in plants usually caused by nonstandard means of dispersal? *Ecology*, **84**, 1945–56.

Hillebrand, H., Watermann, F., Karez, R. and Berninger, U.-G. 2001. Differences in species richness patterns between unicellular and multicellular organisms. *Oecologia*, **126**, 114–24.

Hobbs, R.J. and 17 others. 2006. Novel ecosystems: theoretical and management aspects of the new ecological world order. *Global Ecology and Biogeography*, **15**, 1–7.

Hodkinson, I.D. 1975. Dry weight loss and chemical changes in vascular plant litter of terrestrial origin, occurring in a beaver pond ecosystem. *Journal of Ecology*, **63**, 131–42.

Hodkinson, I.D. and Casson, D. 1991. A lesser predilection for bugs: Hemiptera (Insecta) diversity in tropical rain forests. *Biological Journal of the Linnean Society*, **43**, 101–9.

Hodkinson, I.D. and Coulson, S.J. 2004. Are high Arctic terrestrial food chains really that simple?—The Bear Island food web revisited. *Oikos*, **106**, 427–31.

Hodkinson, I.D., Coulson, S.J. and Webb, N.R. 2004. Invertebrate community assembly along proglacial chronosequences in the High Arctic. *Journal of Animal Ecology*, **73**, 556–68.

Hoffman, L.R., D'Argenio, D.A., MacCross, M.J., Zhang, Z, Jones, R.A. and Miller, S.I. 2005. Aminoglycoside antibiotics induce bacterial biofilm formation. *Nature*, **436**, 1171–5.

Holland, J.H. 1992. Genetic algorithms. *Scientific American*, **266**, 44–50.

Holmes, A. 1944. *Principles of Physical Geology*. London: Thomas Nelson.

Hoyle, F. 1950. *The nature of the universe*. Oxford: Basil Blackwell.

Hoyle, F. 1957. *The Black Cloud*. London: William Heinemann.

Hoyle, F. 1964. *Of men and galaxies*. Washington: Washington University Press.

Hoyle, F. and Wickramasinghe, N.C. 1999a. Comets—a vehicle for panspermia. *Astrophysics and Space Science*, **268**, 333–41. [first 'published' in preprint form in 1981].

Hoyle, F. and Wickramasinghe, N.C. 1999b. The Universe and life: deductions from the weak anthropic principle. *Astrophysics and Space Science*, **268**, 89–102.

Huggett, R.J. 1999. Ecosystem, biosphere, or Gaia? What to call the global ecosystem. *Global Ecology and Biogeography*, **8**, 425–31.

Huntley, B. 1994. Plant species' response to climate change: implications for the conservation of European birds. *Ibis*, **137**, s127–s138.

Huntley, B. and Webb T. 1989. Migration: species' response to climatic variations caused by changes in the Earth's orbit. *Journal of Biogeography*, **16**, 5–19.

Huston, M.A. 1997. Hidden treatments in ecological experiments: re-evaluating the ecosystem function of biodiversity. *Oecologia*, **110**, 449–60.

Huxley, T.H. 1887. *Physiography: an introduction to the study of nature*. 3rd ed. London: MacMillan.

Innes, J.B. and Blackford, J.J. 2003. The ecology of Late Mesolithic woodland disturbances: model testing with fungal spore assemblage data. *Journal of Archaeological Science*, **30**, 185–94.

Janzen, D.H. 1977. Why fruits rot, seeds mold, and meat spoils. *American Naturalist*, **111**, 691–713.

Janzen, D.H. 1979. Why food rots. *Natural History*, **86**, 60–4.

Janzen, D.H. 1986. Lost plants. *Oikos*, **46**, 129–31.

Jepson, P. and Canney, S. 2003. Values-led conservation. *Global Ecology and Biogeography*. **12**, 271–74.

Johnson, N.P. and Mueller, J. 2002. Updating the accounts: global mortality of the 1918–1920 'Spanish' influenza pandemic. *Bulletin of the History of Medicine*, **76**, 105–15.

Jones, C.G., Lawton, J.H. and Shachak, M. 1994. Organisms as ecosystem engineers. *Oikos*, **69**, 373–86.

Jones, C.G., Lawton, J.H. and Shachak, M. 1997. Positive and negative effects of organisms as physical ecosystem engineers. *Ecology*, **78**, 1946–57.

Jones, D. 2005. Personal effects. *Nature*, **438**, 14–16.

Kassen, R., Buckling, A., Bell, G. and Rainey, P.B. 2000. Diversity peaks at intermediate productivity in a laboratory microcosm. *Nature*, **406**, 508–12.

Kassen, R., Llewellyn, M. and Rainey, P.B. 2004. Ecological constraints on diversification in a model adaptive radiation. *Nature*, **431**, 984–8.

Kasting, J.F. and Siefert, J.L. 2002. Life and the evolution of Earth's atmosphere. *Science*, **296**, 1066–8.

Kasting, J.F., Whitmire, D.P. and Reynolds, R.T. 1993. Habitable zones around main sequence stars. *Icarus*, **101**, 108–28.

Kennedy, M.J., Reader, S.L. and Swierczynski, L.M. 1994. Preservation records of micro-organisms: evidence of the tenacity of life. *Microbiology*, **140**, 2513–29.

Kenrick, P. and Crane, P.R. 1997. The origin and early evolution of plants on land. *Nature*, **389**, 33–9.

Keppler, F., Hamilton, J.T.G., Brass, M. and Röckmann, T. 2006. Methane emissions from terrestrial plants under aerobic conditions. *Nature*, **439**, 187–91.

Kibby, G. 1979. *Mushrooms and Toadstools: a field guide*. Oxford: Oxford University Press.

Kingsland, S.E. 1995. *Modelling nature, episodes in the history of population ecology*. 2nd ed. Chicago: Chicago University Press.

Kirchner, J.W. 2002. The Gaia hypothesis: fact, theory, and wishful thinking. *Climatic Change*, **52**, 391–408.

Kirchner, J.W. 2003. The Gaia hypothesis: conjectures and refutations. *Climatic Change*, **58**, 21–45.

Kirkwood, T.B.L. and Austad, S.N. 2000. Why do we age? *Nature*, **408**, 233–8.

Kissel, J. and Kruger, F.R. 1987. The organic component in dust from comet Halley as measured by the PUMA mass spectrometer on board Vega 1. *Nature*, **326**, 755–60.

Kleidon, A. 2002. Testing the effect of life on Earth's functioning: how Gaian is the Earth system? *Climatic Change*, **52**, 383–9.

Kleidon, A. 2004. Beyond Gaia: thermodynamics of life and Earth system functioning. *Climatic Change*, **66**, 271–319.

Klinger, L.F., Taylor, J.A. and Franzen, L.G. 1996. The potential role of peatland dynamics in ice-age initiation. *Quaternary Research*, **45**, 89–92.

Krebs, C.J. 2001. *Ecology*. 5th ed. London: Benjamin Cummings.

Kühlbrandt, W. 2001. Chlorophylls galore. *Nature*, **411**, 896–9.

Kuhlman, K.R. and 10 others. 2005. Enumeration, isolation, and characterization of ultraviolet (UV-C) resistant bacteria from rock varnish in the Whipple Mountains, California. *Icarus*, **174**, 585–95.

Kump, L.R., Kasting, J.F. and Crane, R.C. 2004. *The Earth System*. 2nd ed. New Jersey: Pearson.

Lacy, R.C. 2000. Structure of the VORTEX simulation model for population viability analysis. *Ecological Bulletins*, **48**, 191–203.

Lal, R. 2004. Soil carbon sequestration to mitigate climate change. *Geoderma*, **123**, 1–22.

Laland, K.N., Odling-Smee, F.J. and Feldman, M.W. 1999. Evolutionary consequences of niche construction and their implications for ecology. *Proceedings of the National Academy of Science USA*, **96**, 10242–7.

Lande, R. 1993. Risks of population extinction from demographic and environmental stochasticity and random catastrophes. *American Naturalist*, **142**, 911–27.

Lange, O.L., Kidron, G.J., Budel, B., Meyer, A., Kilian, E. and Abeliovich, A. 1992. Taxonomic composition and photosynthetic characteristics of the 'biological soil crusts' covering sand dunes in the western Negev Desert. *Functional Ecology*, **6**, 519–27.

Lawton, J. 2001. Earth System Science. *Science*, **292**, 1965.

Lawton, J.H. 1996. Patterns in ecology. *Oikos*, **75**, 145–7.

Lawton, J.H. 1999. Are there general laws in ecology? *Oikos*, **84**, 177–92.

Leakey, R. and Lewin, R. 1995. *The sixth extinction: biodiversity and its survival.* London: Weidenfeld and Nicolson.

Lee, D.W., O'Keefe, J., Holbrook, N.M. and Feild, T.S. 2003. Pigment dynamics and autumn leaf senescence in a New England deciduous forest, eastern USA. *Ecological Research*, **18**, 677–94.

Lenton, T.M. 1998. Gaia and natural selection. *Nature*, **394**, 439–47.

Lenton, T.M. 2001. The role of land plants, phosphorus weathering and fire in the rise and regulation of atmospheric oxygen. *Global Change Biology*, **7**, 613–29.

Lenton, T.M. 2004. Clarifying Gaia: regulation with or without Natural Selection. In *Scientists debate Gaia.* (ed. Schneider, S.H., Miller, J.R., Crist, E. and Boston, P.J.) pp. 15–25.Cambridge Massachusetts: MIT Press.

Lenton, T.M. and Huntingford, C. 2003. Global terrestrial carbon storage and uncertainties in its temperature sensitivity examined with a simple model. *Global Change Biology*, **9**, 1333–52.

Lenton, T.M. and Lovelock, J.E. 2001. Daisyworld revisited: quantifying biological effects on planetary self-regulation. *Tellus*, **53B**, 288–305.

Lenton, T.M., Schellnhuber, H.J. and Szathmáry, E. 2004. Co-evolution: Earth history involves tightly entwined transitions of information and the environment, but where is this process heading? *Nature*, **431**, 913.

Lenton, T.M. and Watson, A.J. 2000a. Redfield revisited 1. Regulation of nitrate, phosphate, and oxygen in the ocean. *Global Biogeochemical Cycles*, **14**, 225–48.

Lenton, T.M. and Watson, A.J. 2000b. Redfield revisited 2. What regulates the oxygen content of the atmosphere? *Global Biogeochemical Cycles*, **14**, 249–68.

Lenton T.M. and Watson, A.J. 2004. Biotic enhancement of weathering, atmospheric oxygen and carbon dioxide in the Neoproterozoic. *Geophysical Research Letters*, **31**, L05202, doi:10.1029/2003GLO18802.

Lenton, T.M. and Wilkinson, D.M. 2003. Developing the Gaia theory, a response to the criticisms of Kirchner and Volk. *Climatic Change*, **58**, 1–12.

Lepland, A., van Zuilen, M.A., Arrhenius, G., Whitehouse, M.J. and Fedo, C.M. 2005. Questioning the evidence for Earth's earliest life—Akilia revisited. *Geology*, **33**, 77–9.

Lever, C. 1977. *The naturalized animals of the British Isles.* London: Hutchinson.

Lev-Yadum, S., Gopher, A. and Abbo, S. 2000. The cradle of agriculture. *Science*, **288**, 1602–3.

Lewontin, R. 2000. *The triple helix.* Cambridge Mass: Harvard University Press.

Lindbladh, M. and Bradshaw, R. 1995. The development and demise of a medieval forest-meadow system at Linnaeus' birthplace in southern Sweden: implications for conservation and forest history. *Vegetation History and Archaeobotany*, **4**, 153–60.

Lipman, C.B. 1928. The discovery of living microorganisms in ancient rocks. *Science*, **68**, 272–31.

Lomax, B., Beerling, D., Upcherch Jr., G. and Otto-Bliesner, B. 2001. Rapid (10-yr) recovery of terrestrial productivity in a simulation study of the terminal Cretaceous impact event. *Earth and Planetary Science Letters*, **192**, 137–44.

Lorenz, R. 2003. Full steam ahead-probably. *Science*, **299**, 837–8.

Lorenz, R.D. 2002. Planets, life and the production of entropy. *International Journal of Astrobiology*, **1**, 3–13.

Lorenz, R.D., Lunine, J.I., Withers, P.G. and McKay, C.P. 2001. Titan, Mars and Earth: entropy production by latitudinal heat transport. *Geophysical Research Letters*, **28**, 415–18.

Lovelock, J. 1979. *Gaia*. Oxford: Oxford University Press.

Lovelock, J. 2000a. *The ages of Gaia*. 2nd ed. Oxford: Oxford University Press.

Lovelock, J. 2000b. *Homage to Gaia*. Oxford: Oxford University Press.

Lovelock, J. 2003. The living Earth. *Nature*, **426**, 769–70.

Lovelock J. 2006. *The revenge of Gaia*. London: Alan Lane.

Lovelock, J.E. 1965. A physical basis for life detection experiments. *Nature*, **207**, 568–70.

Lovelock, J.E. 1992. A numerical model for biodiversity. *Philosophical Transactions of the Royal Society B*, **338**, 383–91.

Lovelock, J.E., Maggs, R.J. and Wade, R.J. 1973. Halogenated Hydrocarbons in and over the Atlantic. *Nature*, **241**, 194–6.

Lovelock, J.E. and Margulis, L. 1974. Atmospheric homeostasis by and for the biosphere: the gaia hypothesis. *Tellus*, **16**, 2–10.

Lovelock, J.E. and Watson, A.J. 1982. The regulation of carbon dioxide and climate: Gaia or geochemistry. *Planet Space Science*, **30**, 795–802.

Macdonald, D.W., Tattrsall, F.H., Brown, E.D. and Balharry, D. 1995. Reintroducing the European Beaver to Britain: nostalgic meddling or restoring biodiversity? *Mammal Review*, **25**, 161–200.

MacMahon, J.A. 1985. *Deserts*. New York: Alfred A. Knopf.

Maddock, L. 1991. Effects of simple environmental feedback on some population models. *Tellus*, **43B**, 331–7.

Madigan, M.T., Martinko, J.M. and Parker, J. 2000. *Brock: Biology of microorganisms*. 9th ed. New Jersey: Prentice Hall.

Maitland, P.S. and Campbell, R.N. 1992. *Freshwater fishes*. London: Harper Collins.

Majerus, M.E.N. 1994. *Ladybirds*. London: Collins.

Majerus, M.E.N. 2002. *Moths*. London: Harper Collins.

Margulis, L. 1971. The origin of plant and animal cells. *American Scientist*, **59**, 230–5.

Margulis, L. and Dolan, M.F. 2002. *Early life*. 2nd ed. Boston: Jones and Bartlett.

Margulis, L. and Lovelock, J.E. 1974. Biological modulation of the Earth's atmosphere. *Icarus*, **21**, 471–89.

Margulis, L. and Sagan, D. 1995. *What is life*? London: Wiedenfeld and Nicolson.

Margulis, L. and Sagan, D. 1997. *Slanted truths*. New York: Copernicus.

Margulis, L. and Schwartz, K.V. 1998. *Five Kingdoms* 3rd ed. New York: W.H. Freeman.

May, R.M. 1973. *Stability and complexity in model ecosystems*. Princeton: Princeton University Press.

May, R.M. 1974. Biological populations with non-overlapping generations: stable points, stable cycles, and chaos. *Science*, **186**, 645–7.

May, R.M. 1988. How many species are there on Earth? *Science*, **241**, 1441–9.

Maynard Smith, J. and Szathmary, E. 1995. *The major transitions in evolution*. Oxford: W.H. Freeman.

Mayr, E. 1982. *The growth of biological thought*. Cambridge, Mass: Harvard University Press.

Mayr, E. 2004. *What makes biology unique?* Cambridge: Cambridge University Press.

Mayr, O. 1986. *Authority, Liberty and automatic machinery in early modern Europe*. Baltimore: John Hopkins University Press.

McCallum, H., Kikkawa, J. and Catterall, C. 2000. Density dependence in an island population of silvereyes. *Ecology Letters*, **3**, 95–100.

Merilaita, S., Lyytinen, A. and Mappes, J. 2001. Selection for cryptic coloration in a visually heterogeneous habitat. *Proceedings of the Royal Society B.*, **268**, 1925–9.

Merilaita, S., Tuomi, J. and Jormalainen, V. 1999. Optimization of cryptic coloration in heterogeneous habitats. *Biological Journal of the Linnean Society*, **67**, 151–61.

Meshik, A.P. 2005. The workings of an ancient nuclear reactor. *Scientific American*, **293**, 56–63.

Midgley, M. 1989. *Wisdom, information and wonder: what is knowledge for?* London: Routledge.

Midgley, M. 2001. *Science and Poetry*. London: Routledge.

Miller, J. 1978. *The body in question*. London: Jonathan Cape.

Miller, P.S. and Lacy, R.C. 2003. *VORTEX: A stochastic simulation of the extinction process. Version 9.21 users manual*. Apple Valley MN: Conservation Specialist Group (SSC/IUCN).

Miskin, I., Rhodes, G., Lawlor, K., Saunders, J.R. and Pickup, R.W. 1998. Bacteria in post-glacial freshwater sediments. *Microbiology*, **144**, 2427–39.

Monteith, J.L. 1981. Evaporation and surface temperature. *Quarterly journal of the Royal Meteorological Society*, **107**, 1–27.

Monteith, J.L. and Unsworth, M.H. 1990. *Principles of environmental physics*. 2nd ed. London: Edward Arnold.

Mooney, H.A. and Ehleringer J.R. 1997. Photosynthesis. In *Plant Ecology* 2nd ed. (ed. Crawley M.) pp. 1–27. Oxford: Blackwell.

Moorbath, S. 2005. Dating earliest life. *Nature*, **434**, 155.

Moore, P.D. 1975. Origin of blanket mires. *Nature*, **256**, 267–9.

Moore, P.D. 1993. The origin of blanket mire, revisited. In *Climate change and human impact on the landscape*. (ed. Chambers, F.M.) pp. 217–24. London: Chapman and Hall.

Moore, P.D. 2001. A never-ending story. *Nature*, **409**, 565.

Moore, P.D. 2002. The future of cool temperate bogs. *Environmental Conservation*, **29**, 3–20.

Morton, O. 2002. *Mapping Mars*. London: Forth Estate.

Moss, B. 2001. *The Broads*. London: Harper Collins.

Muir, R. 2004. *Landscape encyclopaedia*. Macclesfield: Windgather.

Mulder, C., Cohen, J.E., Setala, Bloem, J. and Breure, A.M. 2005. Bacterial traits, organism mass, and numerical abundance in detrital soil food web of Dutch agricultural grasslands. *Ecology Letters*, **8**, 80–90.

Myers, N. and Kent, J. 2003. New consumers: The influence of affluence on the environment. *Proceedings of the National Academy of Sciences USA*, **100**, 4963–8.

Naeem, S. 1998. Species redundancy and ecosystem reliability. *Conservation Biology*, **12**, 39–45.

Naeem, S., Thompson, L.J., Lawler, S.P., Lawton, J.H. and Woodfin, R.M. 1994. Declining biodiversity can alter the performance of ecosystems. *Nature*, **368**, 734–7.

Narlikar, J.V. 1999. *Seven wonders of the cosmos*. Cambridge: Cambridge University Press.

Newmark, W.D. 1987. A Land-bridge island perspective on mammalian extinctions in western North American parks. *Nature*, **325**, 430–2.

Nisbet, E.G. 1991. *Leaving Eden; To protect and manage the Earth*. Cambridge: Cambridge University Press.

Nisbet, E.G. 2001. Where did early life live, and what was it like? In *Earth Systems Science. A new subject for study (Geophysiology) or a new philosophy?* (ed. Guerzoni, S., Harding, S., Lenton, T. and Ricci Lucchi, F.) pp. 43–52. Siena: International School Earth and Planetary Science.

Nisbet, E.G. and Fowler, C.M.R. 2003. The early history of life. *Treatise on Geochemistry*, **8**, 1–39.

Nisbet, E.G. and Sleep, N.H. 2001. The habitat and nature of early life. *Nature*, **409**, 1083–91.

Nobis, M. and Wohlgemuth, T. 2004. Trend words in ecological core journals over the last 25 years (1978–2002). *Oikos*, **106**, 411–21.

Novotny, V. and Basset, Y. 2005. Host specificity of insect herbivores in tropical forests. *Proceedings of the Royal Society B*, **272**, 1083–90.

O'Brien, D.P., Lorenz, R.D. and Lunine, J.I. 2005. Numerical calculations on the longevity of impact oases on Titan. *Icarus*, **173**, 243–53.

Odling-Smee, F.J., Laland, K.N. and Feldman, M.W. 1996. Niche construction. *American Naturalist*, **147**, 641–8.

Odling-Smee, F.J., Laland, K.N. and Feldman, M.W. 2003. *Niche Construction*. Princeton: Princeton University Press.

Odum, E.P. 1969. The strategy of ecosystem development. *Science*, **164**, 262–70.

O'Regan, H.J. and Turner, A. 2004. Biostratigraphic and palaeoecological implications of new fossil felid material from the Plio-Pleistocene site of Tegelen, The Netherlands. *Palaeontology*, **47**, 1181–93.

O'Regan, H.J., Turner, A. and Wilkinson, D.M. 2002. European Quaternary refugia: a factor in large carnivore extinction? *Journal of Quaternary Science*, **17**, 789–95.

Osterholm, M.T. 2005. A weapon the world needs. *Nature*, **435**, 417–18.

Ozawa, H. and Ohmura, A. 1997. Thermodynamics of a global-mean state of the atmosphere—a state of maximum entropy increase. *Journal of Climate*, **10**, 441–45.

Pace, N.R. 2001. The universal nature of biochemistry. *Proceedings of the National Acadamy of Sciences USA*, **98**, 805–8.

Paltridge, G.W. 1975. Global dynamics and climate change: a system on minimum entropy exchange. *Quarterly Journal of the Royal Meteorological Society*, **101**, 475–84.

Parker, A. 2005. *Seven deadly colours*. London: The Free Press.

Parker, A.R. 1998. Colour in Burgess Shale animals and the effect of light on evolution in the Cambrian. *Proceedings of the Royal Society B*, **265**, 967–72.

Parks, S.A. and Harcourt, A.H. 2002. Reserve size, local human density, and mammalian extinctions in U.S. protected areas. *Conservation Biology*, **16**, 800–8.

Patenaude, G.L., Briggs, B.D.J., Milne, R., Rowland, C.S., Dawson, T.P. and Pryor, S.N. 2003. The carbon pool in a British semi-natural woodland. *Forestry*, **75**, 109–19.

Pegg, D.E. 2002. The history and principles of cryopreservation. *Seminars in Reproductive Medicine*, **20**, 5–13.

Penrose, R. 2004. *The road to reality*. London: Jonathan Cape.

Perrin, J. 1990. *Yes to dance*. Sparkford: Oxford Illustrated Press.

Perutz, M.F. 1987. Physics and the riddle of life. *Nature*, **326**, 555–8.

Pielou, E.C. 2001. *The energy of nature*. Chicago: Chicago University Press.

Pillans, B. and Naish, T. 2004. Defining the Quaternary. *Quaternary Science Reviews*, **23**, 2271–82.

Pimm, S.L. 1991. *The balance of nature*? Chicago: University of Chicago Press.

Pitelka, L.F. and the Plant Migration Workshop Group. 1997. Plant Migration and Climate Change. *American Scientist*, **85**, 464–73.

Polis, G.A. 1999. Why are parts of the world green? Multiple factors control productivity and the distribution of biomass. *Oikos*, **86**, 3–15.

Pope, K.O., D'Hondt, S.L. and Marshall, C.R. 1998. Meteorite impact and the mass extinction of species at the Cretaceous/Tertiary boundary. *Proceedings of the National Academy of Sciences USA*, **95**, 11028–9.

Porley, R. and Hodgetts, N. 2005. *Mosses and Liverworts*. London: Harper Collins.

Porter, R. 1997. *The greatest benefit to mankind*. London: Harper Collins.

Porter, R. 2000. *Enlightenment*. London: Allen Lane.

Postgate, J. 1990. The microbes that would not die. *New Scientist*, **127**, 46–9.

Postgate, J. 2000. *Microbes and man*. 4th ed. Cambridge: Cambridge University Press.

Pounds, J.A. and 13 others. 2006. Widespread amphibian extinctions from epidemic disease driven by global warming. *Nature*, **439**, 161–7.

Preston, C.D., Pearman, D.A. and Dines, T.D. 2002. *New Atlas of the British and Irish Flora*. Oxford: Oxford University Press.

Pujol. T. 2002. The consequence of maximum thermodynamic efficiency in Daisyworld. *Journal of Theoretical Biology*, **217**, 53–60.

Purvis, A. and Hector, A. 2000. Getting the measure of biodiversity. *Nature*, **405**, 212–19.

Quammen, D. 1996. *The song of the dodo*. London: Random House.

Rackham, O. 1986. *The history of the countryside*. London: Dent.

Rackham, O. 2003. *Ancient woodland* 2nd ed. Colvend: Castlepoint Press.

Rainey, P.B. and Travisano, M. 1998. Adaptive radiation in a heterogenous environment. *Nature*, **394**, 69–72.

Ramsbottom, J. 1953. *Mushrooms and Toadstools*. London: Collins.

Raven, J.A. 1998. The twelfth Tansley lecture. Small is beautiful: the picophytoplankton. *Functional Ecology* **12**, 503–13.

Raven, P.H. and Johnson, G.B. 1999. *Biology* 5th ed. Boston: McGraw-Hill.

Redfield, A.C. 1958. The biological control of the chemical factors in the environment. *American Scientist*, **46**, 205–21.

Retallack, G.J. 1997. Early forest soils and their role in Devonian global change. *Science*, **276**, 583–5.

Retallack, G.J. 2000. Ordovician life on land and early Palaeozoic global change. *The Palaeontological Society Papers*, **6**, 21–45.

Retallack, G.J. 2001. *Soils of the past*. 2nd ed. Oxford: Blackwell Science.

Rickleffs, R.E. and Miller, G.L. 2000. *Ecology* 4th ed. New York: W.H. Freeman.

Ridgwell, A. and Zeebe, R.E. 2005. The role of the global carbonate cycle in the regulation and evolution of the Earth system. *Earth and Planetary Science Letters*, **234**, 299–315.

Riebesell, U., Zondervan, I., Rost, B., Tortell, P.D., Zeebe, R.E. and Morel, F.M.M. 2000. Reduced calcification of marine plankton in response to increased atmospheric CO_2. *Nature*, **407**, 364–7.

Rigby, S. and Milsom, C.V. 2000. Origins, evolution, and diversification of zooplankton. *Annual Reviews of Ecology and Systematics*, **31**, 293–313.

Robinson, J.M. 1990. Lignin, land plants, and fungi: Biological evolution affecting Phanerozoic oxygen balance. *Geology*, **15**, 607–10.

Robinson, J.M. 1991. Fire in Phanerozoic Cybernetics. In *Scientists on Gaia*. (ed. Schneider, S.H. and Boston, P.J.) pp. 363–72. Cambridge Massachusetts: MIT Press.

Rocap, G and 23 others. 2003. Genome divergence in two *Prochlorococcus* ecotypes reflects oceanic niche differentiation. *Nature*, **424**, 1042–7.

Rodwell, J.S. (ed.) 1991. *British Plant Communities Vol 2. Mires and Heaths*. Cambridge: Cambridge University Press.

Rosell, F., Bozsér, O., Collen, P. and Parker, H. 2005. Ecological impact of beavers Castor fiber and Castor canadensis and their ability to modify ecosystems. *Mammal Review*, **35**, 248–76.

Rosing, M.T. 1999. ^{13}C-Depleted carbon microparticles in >3700-Ma sea-floor sedimentary rocks from West Greenland. *Science*, **283**, 674–6.

Rothschild, L.J. and Mancinelli, R.L. 2001. Life in extreme environments. *Nature*, **409**, 1092–101.

Rowantree, R.A. and Nowak, D.J. 1991. Quantifying the role of urban forests in removing atmospheric carbon dioxide. *Journal of Arboriculture*, **17**, 269–75.

Ruddiman, W.F. 2003. The anthropic greenhouse era began thousands of years ago. *Climatic Change*, **61**, 261–93.

Ruddiman, W.F. 2005. *Plows, plagues and Petroleum*. Princeton: Princeton University Press.

Ruxton, G.D., Sherratt, T.N. and Speed, M.P. 2004. *Avoiding Attack: The evolutionary ecology of crypsis, warning signals and mimicry*. Oxford: Oxford University Press.

Sagan, C. and Chyba, C. 1997. The early faint sun paradox: organic shielding of ultraviolet-labile greenhouse gasses. *Science*, **276**, 1217–21.

Sagan, C., Thompson, W.R., Carlson, R., Gurnett, D. and Hord, C. 1993. A search for life on Earth from the Galileo spacecraft. *Nature*, **365**, 715–21.

Santini, F. and Galleni, L. 2001. Non-light based ecosystems and bioastronomy. *Rivista di Biologia*, **94**, 427–42.

Saunders, P.T. 1994. Evolution without natural selection: further implications of the Daisyworld parable. *Journal of Theoretical Biology*, **166**, 365–73.

Schama, S. 2000. *A history of Britain Vol 1. 300BC–1603 AD*. London: BBC.

Schindler, T.L. and Kasting, J.F. 2000. Synthetic spectra of simulated terrestrial atmospheres containing possible biomarker gases. *Icarus*, **145**, 262–71.

Schneider, S.H., Miller, J.R., Crist, E. and Boston, P.J. (eds) 2004. *Scientists debate Gaia: the next century*. Cambridge, Mass: MIT Press.

Schopf, J.W. 1999. *Cradle of life, the discovery of Earth's earliest fossils*. Princeton, Princeton University Press.

Schopf, J.W. 2000. The fossil record: tracing the roots of the cyanobacterial lineage. In *The Ecology of cyanobacteria*. (ed. Whitton, B.A. and Potts, M.) pp. 13–35. Dordrecht: Kluwer.

Schrodinger, E. 1948. *What is life?* Cambridge: Cambridge University Press.

Schwartzman, D. 1999. *Life, temperature, and the Earth*. New York: Columbia University Press.

Schwartzman, D.W. and Volk, T. 1989. Biotic enhancement of weathering and the habitability of Earth. *Nature*, **340**, 457–60.

Selden, P.A., Corronca, J.A. and Hünicken, M.A. 2005. The true identity of the supposed giant fossil spider Megarachne. *Biology Letters*, **1**, 44–8.

Seto, M. and Akagi, T. 2005. Daisyworld inhabited with daisies incorporating a seed size/number trade off: the mechanism of negative feedback on selection from a standpoint of the competition theory. *Journal of Theoretical Biology*, **234**, 167–72.

Sherratt, T.N., Wilkinson, D.M. and Bain, R.S. 2006. Why fruits rot, seeds mold and meat spoils: a reappraisal. *Ecological Modelling*. **192**, 618–626.

Shklovskii, I.S. and Sagan, C. 1977. *Intelligent life in the universe*. London: Picador.

Simberloff, D. 1998a. Small and declining populations. In *Conservation Science and action*. (ed. Sutherland, W.J.) pp. 116–34. Oxford: Blackwell.

Simberloff, D. 1998b. Flagships, umbrellas, and keystones: is single-species management passé in the landscape era? *Biological Conservation*, **83**, 247–57.

Simo, R. 2001. Production of atmospheric sulfur by oceanic plankton: biogeochemical, ecological and evolutionary links. *Trends in Ecology and Evolution*, **16**, 287–94.

Sinclair, A.R.E. 1989. Population regulation in animals. In *Ecological concepts*. (ed. Cherratt J.M.) pp. 197–241. Oxford: Blackwell.

Sleep, N., Zahnle, K.J., Kasting, J.F. and Marowitz, H.J. 1989. Annihilation of ecosystems by large asteroid impacts on the early Earth. *Nature* **342**, 139–142.

Slobodkin, L.B. and Slack, N.G. 1999. George Evelyn Hutchinson: 20th-century ecologist. *Endeavour*, **23**, 24–30.

Smith, R.L. and Smith, T.M. 1998. *Elements of ecology* 4th ed. Menlo Park, California: Addison Wesley Longman.

Smolin, L. 1997. *The life of the cosmos*. London: Weidenfeld and Nicolson.

Snow, C.P. 1959. *The two cultures*. Cambridge: Cambridge University Press.

Sommer, C. and Bibby, B.M. 2002. The influence of veterinary medicines on the decomposition of dung organic matter in soil. *European Journal of Soil Biology*, **38**, 155–9.

Sorensen, R. 1991. Thought experiments. *American Scientist*, **79**, 250–63.

Spooner, B. and Roberts, P. 2005. *Fungi*. London: Collins.

Stephens, P.A. and Sutherland, W.J. 1999. Consequences of the Allee effect for behaviour, ecology and conservation. *Trends in Ecology and Evolution*, **14**, 401–5.

Stewart, I. 1998. *Life's other secret*. London: Allen Lane.

Stiling, P. 2002. *Ecology* 4th ed. New Jersey: Prentice Hall.

Stork, N. and Gaston, K. 1990. Counting species one by one. *New Scientist*, **127**, 43–7.

Sugimoto, T. 2002. Darwinian evolution does not rule out the Gaia hypothesis. *Journal of Theoretical Biology*, **218**, 447–55.

Sullivan, M.B., Waterbury, J.B. and Chisholm, S.W. 2003. Cyanophages infecting the oceanic cyanobacterium *Prochlorococcus*. *Nature*, **424**, 1047–51.

Sunda, W., Kieber, D.J., Kiene, R.P. and Huntsman, S. 2002. An antioxidant function for DMSP and DMS in marine algae. *Nature*, **418**, 317–20.

Suttle, C.A. 2005. Viruses in the sea. *Nature*, **437**, 356–61.

Tice, M.M. and Lowe, D.R. 2004. Photosynthetic microbial mats in the 3416-Myr-old ocean. *Nature*, **431**, 549–52.

Tilman, D. 2000. Causes, consequences and ethics of biodiversity. *Nature*, **405**, 208–11.

Tilman, D., Lehman, C.L. and Bristow, C.E. 1998. Diversity-stability relationships: statistical inevitability or ecological consequence? *American Naturalist*, **151**, 277–82.

Tilman, D., Wedin, D. and Knops, J. 1996. Productivity and sustainability influenced by biodiversity in grassland ecosystems. *Nature*, **379**, 718–20.

Tolonen, K., Warner, B.G. and Vasander, H. 1992. Ecology of Testaceans (Protozoa: Rhizopoda) in Mires in Southern Finland: 1. Autecology. *Archiv fur Protistenkunde*, **142**, 119–38.

Tompkins, D.M. and Begon, M. 1999. Parasites can regulate wildlife populations. *Parisitology Today*, **15**, 311–12.

Turner, J., Innes, J.B. and Simmons, I.G. 1993, Spatial diversity in the mid-Flandrian vegetation history of North Gill, North Yorkshire. *New Phytologist*, **123**, 599–647.

Tyrrell, T. 1999. The relative influence of nitrogen and phosphorous on oceanic primary production. *Nature*, **400**, 525–31.

Tyrrell, T. 2004. Biotic plunder: control of the environment by biological exhaustion of resources. In *Scientists debate Gaia*. (ed. Schneider, S.H., Miller, J.R., Crist, E. and Boston, P.J.) pp. 137–47. Cambridge Massachusetts: MIT Press.

Van Damme, R., Wilson, R.S., Vanhooydonck, B. and Aerts, P. 2002. Performance constraints in decathletes. *Nature*, **415**, 755–6.

Vepsäläinen, K. and Spence, J.R. 2000. Generalization in Ecology and Evolutionary Biology: from hypothesis to paradigm. *Biology and Philosophy*, **15**, 211–38.

Vernadsky, V.I. 1998. *The Biosphere*. New York: Copernicus. (first published 1926).

Vincent, W.F. 2000. Cyanobacterial dominance in the polar regions. In *The ecology of cyanobacteria*. (ed. Whitton, B.A. and Potts, M.) pp. 321–40. Dordrecht: Kluwer.

Vitousek, P. 2004. *Nutrient cycling and limitation*. Oxford: Princeton University Press.

Volk, T. 1987. Feedbacks between weathering and atmospheric CO_2 over the last 100 million years. *American Journal of Science*, **287**, 763–79.

Volk, T. 1998. *Gaia's body*. New York: Copernicus.

Volk, T. 2003. Natural selection, Gaia, and inadvertent by-products. *Climatic Change*, **58**, 13–19.

Volk, T. 2004. Gaia is life in a wasteworld of by-products. In *Scientists debate Gaia* (ed. Schneider, S.H., Miller, J.R., Crist, E. and Boston, P.J.) pp. 27–36. Cambridge Mass: MIT Press.

Vreeland, R.H., Rosenzweig, W.D. and Powers, D.W. 2000. Isolation of a 250 million-year-old halotolerant bacterium from a primary salt crystal. *Nature*, **407**, 897–900.

Wakeford, T. 2001. *Liaisons of life*. New York: John Wiley.

Walker, J.C.G., Hays, P.B. and Kasting, J.F. 1981. A negative feedback mechanism for the long-term stabilisation of Earth's surface temperature. *Journal of Geophysical Research*, **86**, 9776–82.

Wall, R. and Strong, L. 1987. Environmental consequences of treating cattle with the antiparasitic drug ivermectin. *Nature*, **327**, 418–21.

Warming, E. 1909. *Oecology of plants*. Oxford: The Clarendon Press.

Watson, A.J. 1999. Coevolution of the Earth's environment and life: Goldilocks, Gaia and the anthropic principle. In *James Hutton—Present and Future*. (ed. Craig, G.Y. and Hull, J.H.) Geological Society, London, Special Publications **150**, 75–88.

Watson, A.J. 2004. Gaia and observer self selection. In *Scientists debate Gaia*. (ed. Schneider, S.H., Miller, J.R., Crist, E. and Boston, P.J.) pp. 201–8. Cambridge Massachusetts: MIT Press.

Watson, A.J., Bakker, D.C.E., Ridgewell, A.J., Boyd, P.W. and Law, C.S. 2000. Effect of iron supply on Southern Ocean CO_2 uptake and implications for glacial atmospheric CO_2. *Nature*, **407**, 730–3.

Watson, A.J. and Lovelock, J.E. 1983. Biological homeostasis of the global environment: the parable of Daisyworld. *Tellus*, **35B**, 284–9.

Watson, A.J., Lovelock, J.E. and Margulis, L. 1978. Methanogenesis, fires and the regulation of atmospheric oxygen. *BioSystems*, **10**, 293–8.

Wells, L.E., Armstrong, J.C. and Gonzalez, G. 2003. Reseeding of early Earth by impacts of returning ejecta during the late heavy bombardment. *Icarus*, **162**, 38–46.

Welsh, D.T. 2000. Ecological significance of compatible solute accumulation by microorganisms: from single cells to global climate. *FMES Microbiology Reviews*, **24**, 263–90.

West, S.A., Lively, C.M. and Read, A.F. 1999. A pluralist approach to sex and recombination. *Journal of Evolutionary Biology*, **12**, 1003–12.

Wheldale, M. 1916. *The anthocyanin pigments of plants*. Cambridge: Cambridge University Press.

Whitfield, J. 2004. Time lords. *Nature*, **429**, 124–5.

Whitman, W.B., Coleman, D.C. and Wiebe, W.J. 1998. Prokaryotes: the unseen majority. *Proceedings of the National Academy of Sciences USA*, **95**, 6578–83.

Wickramasinghe, C. 2005. *A journey with Fred Hoyle*. New Jersey: World Scientific.

Wildman Jr., R.A., Hickey, L.J., Dickinson, M.B., Berner, R.A., Robinson, J.M., Dietrich, M., Essenhigh, R.H. and Wildman, C.B. 2004. Burning of forest materials under late Paleozoic high atmospheric oxygen levels. *Geology*, **32**, 457–60.

Wilkinson, D.M. 1997. Plant colonization: are wind dispersed seeds really dispersed by birds at larger spatial and temporal scales? *Journal of Biogeography*, **24**, 61–5.

Wilkinson, D.M. 1998. Fragments of an entangled bank: do ecologists study most of ecology? *Oikos*, **82**, 393–4.

Wilkinson, D.M. 1999a. Is Gaia conventional ecology? *Oikos*, **84**, 533–6.

Wilkinson, D.M. 1999b. Plants on the move (inside science No. 119). *New Scientist*, **161**, S1–S4.

Wilkinson, D.M. 2000. Running with the Red Queen: reflections on 'Sex versus non-sex versus parasite.' *Oikos*, **91** 589–96.

Wilkinson, D.M. 2001a. What is the upper size limit for cosmopolitan distribution in free-living microorganisms? *Journal of Biogeography*, **28**, 285–91.

Wilkinson, D.M. 2001b. At cross purposes. *Nature*, **412**, 485.

Wilkinson, D.M. 2003. The fundamental processes in ecology: a thought experiment on extraterrestrial biospheres. *Biological Reviews*, **78**, 171–9.

Wilkinson, D.M. 2004a. The parable of Green mountain: Ascension Island, ecosystem construction and ecological fitting. *Journal of Biogeography*, **31**, 1–4.

Wilkinson, D.M. 2004b. Homeostatic Gaia: an ecologists perspective on the possibility of regulation. In *Scientists debate Gaia*. (ed. Schneider, S.H., Miller, J.R., Crist, E. and Boston, P.J.) pp. 71–6. Cambridge Massachusetts: MIT Press.

Wilkinson, D.M. and Davis, S.R. 2005. An introduction to Chat Moss. In *Quaternary of Rossendale Forest and Greater Manchester*. (ed. Croft, R.G.) pp. 38–43. London: Quaternary Research Association.

Wilkinson, D.M. and O'Regan, H.J. 2003. Modelling differential extinctions to understand big cat distribution on Indonesian Islands. *Global Ecology and Biogeography*, **12**, 519–24.

Wilkinson, D.M. and Sherratt, T.N. 2001. Horizontally acquired mutualisms, an unsolved problem in ecology? *Oikos*, **92**, 377–84.

Wilkinson, D.M., Sherratt, T.N., Phillip, D.M., Wratten, S.D., Dixon, A.G.F. and Young, A.J. 2002. The adaptive significance of autumn leaf colours. *Oikos*, **99**, 402–7.

Williams, G.C. 1992. Gaia, nature worship and biocentric fallacies. *Quarterly Review of Biology*, **67**, 479–86.

Willis, K.J., Gillson, L. and Brncic, T.M. 2004. How 'virgin' is virgin rainforest? *Science*, **304**, 402–3.

Wilson, E.O. 1992. *The diversity of life*. Cambridge, Mass: Harvard University Press.

Wilson, E.O. 2002. *The future of life*. London: Little Brown.

Wolbach, W.S., Gilmour, I., Anders, E., Orth, C.J. and Brooks, R.R. 1988. Global fire at the Cretaceous-Tertiary boundary. *Nature*, **334**, 665–9.

Wolstencroft, R.D. and Raven, J.A. 2002. Photosynthesis: likelihood of occurrence and possible detection on Earth-like planets. *Icarus*, **157**, 535–48.

Wynn-Williams, D.D. 1996. Antarctic microbial diversity: the basis of polar ecosystem processes. *Biodiversity and Conservation*, **5**, 1271–93.

Xiong, J., Fischer, W.M., Inoue, K., Nakahara, M. and Bauer, C.E. 2000. Molecular evidence for the early evolution of photosynthesis. *Science*, **289**, 1724–30.

Yu, D.W. 2001. Parasites of mutualisms. *Biological Journal of the Linnean Society*, **72**, 529–46.

Zhang, X., Shu, D., Li, Y. and Han, J. 2001. New sites of Chengjiang fossils: crucial windows on the Cambrian explosion. *Journal of the Geological Society, London*, **158**, 211–18.

Ziman, J. 2003. Emerging out of nature into history: the plurality of the sciences. *Philosophical Transactions of the Royal Society London. A*, **361**, 1617–33.

Index

Note: page numbers in *italics* refer to Figures and Tables. Those in **bold** refer to Glossary entries.